FLIGHTDECK PROSE & CONS

FLIGHTDECK PROSE & CONS

FLIGHTDECK PROSE & CONS

Secrets from the little room at the front of the plane

By Captain James McBride

an imprint of

flightsofpassion.com

Published by: flightsofpassion.com

www.flightsofpassion.com

@McBrideCaptain

captjamesmcbride

"Last Day in the Office" – YouTube Channel

First published 2021

PAPERBACK ISBN: 978-0-9931368-3-2

Front Cover Photo: Hannah McBride

For:

Hannah & Lawrence

Being your Dad has been the coolest job I ever had!

Also by James McBride

The Flightdeck Survival Manual

Cooking on the Piste

From the Flightdeck

Aircrew Stories Volume 1

This book combines new stories with some previously published in Flight Training News, Flyer Magazine, Flight International or Aeropodium Expert

The author has written regularly for aviation industry publications, websites, and social media since 2006

INTRODUCTION

'Flightdeck Prose & Cons' is a collection of aviation short stories. It is my fourth such book and I very much enjoyed putting it together. As the name suggests there are lots of pros and cons involved in being a pilot. For sure it's not 'all beer and skittles' as the older generation used to say. At times it is potentially hazardous, if not directly to your health, then certainly to your career. Commercial Aviators are rigorously tested every six months in the simulator to ensure their skill sets remain sharp. In addition, they undergo regular extensive medical tests to confirm their fitness to operate increasingly complicated flying machines in a stressful environment.

Of course, it is right that it should be this way. The lives of the travelling public are literally in the hands of the men and women who are sitting in that little room at the front of the plane. Cockpit or Flightdeck, call it what you will, it is the same place. These days the flightdeck door is securely locked and nobody is allowed entry apart from bona fide crew members, In the days prior to 11th September 2001, it was with genuine pleasure we were able to meet the customers face to face. We welcomed them into our office and I like to think they got a better appreciation of what we do in there from their visit.

It is not just that we have a fabulous view from the office window – that happens on the pretty weather days. The view can also be hugely challenging, for example when you are on final approach and looking at the landing runway effectively through a side window due to the crosswind. Or when there is so much low-level turbulence, with gusty winds on finals that the autopilot cannot cope anymore and we fly manually. It is one of the rare times when the human can out-perform the automatics. Much of our flying is automated not only because it is tiring to continuously carry out manual instrument flying, but because the fuel burn is less. This is an important consideration when you are hoping that your company will continue to operate in the future.

✈

Aviation and flying aeroplanes for a living, has held my attention for decades. Firstly, as a young boy gawping at aeroplanes with my Dad and listening to his stories from when he was in the Royal Air Force during the Second World War. Then later. I got my first job and started to learn to fly at the age of 18. My employment as a student nurse gave me £80 per month and flying lessons cost almost £20 per hour. Even I could tell it was going to take a long time to get my

Private Pilots Licence - I had to buy food and other stuff.

Eventually I cracked the code and found a way to fly. I joined the military and never looked back. Like many other aspiring pilots, I broke through a lot of barriers to make a professional career in aviation. My story is not unusual, but some of the things I have witnessed in flying have been. People sometimes ask me, “How do you think of more things to write about?” and I reply, “Well... it’s a big story... and it’s not over yet.” It seems that every time I get a new job with an airline, the buzz starts all over again. I get repeated flashbacks to previous times. Flying comes to me in my dreams even now. Usually there is some form of drama or problem – there are never easy flights in dreams. It’s a tech issue, a passenger problem, or a crew member who is sick, injured or misbehaving. Yes, there’s been plenty of that down the years.

One thing; it has never been dull, but loaded with laughter, surprises, fun and games in the air and on the ground. You never know what’s coming next. I feel blessed to have had the opportunity to meet and work with some of the best people on the planet. I am privileged to have played a part in making other folks’ dreams come true. By that I mean, not just the crew, but the customers. Aviation has brought families and loved ones together across continents. It has carried life-saving medicines and equipment to remote areas and some not so remote. We should all be grateful for the myriad of ways that aviation benefits mankind

Surprises? Like the time I was flying long-haul for Air India out of Mumbai with local mainline cabin crew onboard. Only the pilots up front were Europeans. The senior flight attendant came into the flightdeck once we reached high altitude cruise and introduced herself. Recognising her name was difficult to pronounce for our western tongues, she said, “Call me by my nickname: call me ‘Jugs’!” we pilots looked at each other. I pursed my lips, slowly shook my head and thought, *where - do we go from here?*

On a serious note, some of the chapters contained in this book might appear unconnected. To help put them into context, I have added notes giving the backstories to the events. All my tales are either entirely true or based upon real events. In some cases, the names have been changed to protect the innocent. Read on... and learn some of the secrets from that little room at the front of the plane.

James McBride   11th March 2021

ACKNOWLEDGMENTS

My grateful thanks are due to the following people:

Rosemary Macray and Ashley Wilkes for their diligent copy editing. Ian Seager (Seager publishing and FLYER Magazine) for permission to republish 'Flying the Perfect ILS'. Dominic Perry (Flight Global) for permission to republish 'It's not about the Money'. Capt. Zoe Ebrey for her photo. Jeremy Pratt for his editing advice & support. Capt. Tony Storey for his past mentoring and photo. Alex Hodgson (Hangar 9 Film Company) for book covers and all his great ideas. Capt. Malcolm Kavanagh for his photo and memories from 1988. Panagiotis Panagopoulos (CEO Aeropodium) for his kind support and permission to republish articles from Aeropodium Expert.

"... Rolls Royce RB211 engine, I have never even seen one cough!"
Which used to be entirely true.

CONTENTS

ACKNOWLEDGMENTS

My grateful thanks are due to the following people:

Rosemary Macray and Ashley Wilkes for their diligent copy editing. Ian Seager (Seager publishing and FLYER Magazine) for permission to republish 'Flying the Perfect ILS'. Dominic Perry (Flight Global) for permission to republish 'It's not about the Money'. Capt. Zoe Ebrey for her photo. Jeremy Pratt for his editing advice & support. Capt. Tony Storey for his past mentoring and photo. Alex Hodgson (Hangar 9 Film Company) for book covers and all his great ideas. Capt. Malcolm Kavanagh for his photo and memories from 1988. Panagiotis Panagopoulos (CEO Aeropodium) for his kind support and permission to republish articles from Aeropodium Expert.

1 A LUCKY ENGINE FAILURE

I am sometimes asked by people, "What's the worst thing that's happened to you in your airline career?" with the expectation that I have something particularly ghoulish to share. After all, when you've been in the game for over three decades, you would think there could have been a lot of dramas. I confess that I usually try to get away with a flippant comment such as... "Well... there was a time... when the salary was late!" You see, the industry has a phenomenal safety record as a system for transporting people and materiel all over the planet. It is not my place to harm that reputation in any way by relating anecdotally, events which are extremely rare, such as experiencing an engine failure in a modern jet airliner. These engines are ultra-reliable. I used to say that,

"In all the years and thousands of hours I have been operating the Boeing 757 equipped with Rolls Royce RB211 engines, I have never even seen one cough!" Which used to be entirely true.

Perhaps the only exception to that was on a trip to the Caribbean, inbound to Jamaica when we observed (and felt) abnormally high vibration on one of the engines while descending over eastern Cuba. It was troublesome enough for us to consider the possibility of diverting to another airfield if we were forced to shut it down. Fortunately, before we had to choose between Haiti or Cuba (imagine...?!) the vibration cleared, and the engine decided to behave normally. We considered that we might have picked up some engine ice while passing through cloud, although that would not have explained why it was not symmetrical – the other motor was smooth throughout. After landing in Kingston, we had the engineering team look at it, but they could find nothing wrong at all and declared the machine serviceable.

The RB211 high-bypass turbofan engine was the most reliable of Rolls Royce's products for decades. In fact, the manufacturer at one stage were troubled that because of the reliability, they were not selling enough spares! Developed originally for the Lockheed L1011, it was subsequently fitted to the Boeing 747, 757 and 767. It was even used to power the Tupolev TU-204. The last engine was produced in 1997 and was followed by the Trent Series that owed much of its design features to its predecessor. In the early 1970s, Rolls predicted that the RB211 would have a useful lifespan of 50 years with continuous development and improvement Even though the pundits from Derby miscalculated by 20 years, the RB211 can take credit for putting Rolls Royce Aero-engines on to the global stage.

So, it is with that impeccable background of faultless service that we approached the operation of the aeroplane. Naturally we experienced all the worst possible disasters during the regular six-monthly Simulator checks, but as for flying the line, we never doubted the robustness of the powerplants. On reflection, we should have considered this aspect of the operation a threat. By that I mean, in the event of a serious engine malfunction, we could lose a valuable second or two as our startle factor was intensified by disbelief! So it was, when finally aboard a B757, I experienced a significant engine malfunction inflight, where we had to consider shutting down the motor.

Our duty commenced in the Middle East where we picked up the flight as the relief crew to operate through the night to the UK. One complication was that the flight was a VIP flight with internationally famous people onboard. After an extended turnround at the VIP terminal in Abu Dhabi, we finally managed to get them all back onboard. As you may surmise, it is not an easy job trying to hurry up VIPs – you always find yourself walking a tightrope, balancing the schedule versus protocol and etiquette. It is a no-win situation as a VIP Captain, because if you arrive late, everybody blames you anyway, even if the customers were dawdling during boarding. As a very old Skipper once said to me, "James, it's all reflected in the salary..." – as if that was supposed to make me feel better.

Pushing back a few minutes late I was conscious of the fact that we could make up some time enroute. I indicated to the crew-chief outside to disconnect his headset and pull the steering bypass pin. Waving from my window on the left side and noting the tug reversing away pulling the towbar, I called for the before-taxi checks. These we completed quickly and soon after we were airborne, racing westwards ahead of the following dawn. Well as dawn travels at twice our flying speed (1000 mph), naturally it was an uneven contest and we were overtaken somewhere in eastern Europe. As we touched down at Liverpool Airport almost exactly on time, and taxied in, it was a good feeling knowing that soon we would have the aircraft to ourselves.

✈

The next flight was an empty sector, ferrying to our maintenance airfield in Southern England – just a cool 35-minute hop with crew only on board. Having said goodbye to our passengers, it was mid-morning local time when we closed the doors and called for engine start. Of course, we were all a little jaded from our night-flight, but I would not say we were overly tired. We had been operating as a heavy crew (with a third pilot) and had all taken inflight rest.

Everything was completely normal from the technical point of view as we approached the runway, albeit that due to the nature of this flight we were all feeling a bit more relaxed. You can't help it. There is a natural tendency to chill-out a little when you are ferrying an empty ship. Of course, you brief some of the obvious threats such as the dramatically different take-off performance with an unladen aircraft.

The Boeing 757 is overpowered at the best of times, even with a 30-tonne payload, but when taking-off empty, she's a rocketship! So it was that as I hit the stopwatch and Auto-throttle EPR button Simultaneously, we got a real push in the back as we accelerated down Runway 27. "EightyKnots-VeeOneRotate-POSITIVERATE!" all came super-quick as I hung on to the racehorse's neck... Then we were talking to the Manchester Controller and negotiating a continuous climb... *"This is one homesick angel!"*

Shortly afterwards as we were established in the climb, passing around seven or eight thousand feet from memory, there was a very loud "BANG!" From the right-hand side, then another... and another "BANG!... BANG! ....BANG!!" They sounded like explosions and it seemed the origin was the right-hand engine?! We were incredulous – this was not supposed to happen with an RB211! Simultaneously the right-hand engine instruments were indicating this was serious. They fluctuated wildly with each successive "BANGGG!!" And the BANGS kept coming. After the initial shock, I asked Pilot Monitoring (also a Captain with the airline and a trusted friend) for "Memory Items for Engine Surge/Stall please, I have the radio!" With confirmation from me, he immediately identified and retarded the affected Thrust Lever and I could feel the loss of thrust. I eased the nose forwards and down to maintain speed, sacrificing climb rate – we had plenty of that anyway.

As I did so, I applied left rudder as smoothly as possible and trimmed it out. I think it was around this time that the flightdeck door opened and the Senior Cabin Crew member demanded in a loud voice, *"Exactly WHAT THE F*** is going on in here!?"* which did make us laugh, albeit we didn't have time to reply. By this time, we were deep into actioning the QRH, starting the APU (just in case) and talking to Air Traffic Control.

Don't misunderstand me, although we were surprised, we acted well as a team and once the initial situation was stabilised – idle thrust meant that the engine stopped surging – we had time to carry out formal decision-making process in accordance with best practice. All those years of Simulator training paid off and even though this was a first for all of us (asymmetric thrust for real in the 757) we acted correctly and with confidence. My overriding thought was this. What a truly LUCKY break!? We had just flown a VIP flight for eight hours through the night halfway across the world without any faults whatsoever and now on the very next sector, we were flying empty to maintenance and we have an engine failure – excellent timing!

Having carefully considered our options, we elected not to divert enroute as this would not have been a safer choice. We were already planned and briefed for a short flight to a major UK airfield, we still had both engines running, even though one was at idle thrust. There were nil abnormal symptoms at idle and the APU was running as an immediate backup in the case of a full shutdown. We carried out a situation briefing* with the cabin crew so that all were in the loop and we managed to assure them that the loud backfiring noises from the right engine were not caused by the pilots "messing about in the cockpit!" We had also noted that there was a concomitant Duct Pressure fluctuation with each surge, so we surmised there was a possible connection. After landing the engineers did indeed find a hard fault with the bleed valve on the engine that had caused the surging.

So that was it. "NBD" as they say. Probably, statistically, the only real engine failure of my career as an operating crew member. I still cannot believe how lucky we were that it occurred on that flight. If it had been the previous one, it would have been a very different story. Imagine; climbing out from Abu Dhabi, at night, with VIP Pax in the cabin, those huge explosive noises from the engine – probably flames too... running the QRH, making the decision to either turnback or divert with ALL of the logistical complications after landing? Certainly, they would never have flown with our company again and the damage to our reputation could have been seriously debilitating financially. But it DIDN'T happen, and they will never know about it. But you do.

✈

*Situation briefing for the cabin crew is usually what we call a NITS briefing. The acronym stands for **N**ature (of the emergency), **I**ntentions (of the cockpit crew), **T**ime (remaining airborne), **S**pecial Instructions (e.g. with a landing gear problem, it MAY collapse on touchdown, therefore Pax Evac may be required).

BACKSTORY: True tale from start to finish. We were SO lucky that day. If the engine failure had occurred coming out of Abu Dhabi, it would have been devastating for our company. The charter brokers of the rich and famous do not book an operator they do not feel 100 percent comfortable with regarding safety and security.

They are incredibly nervous (bordering on neurotic) and sometimes even superstitious. If we had returned to land in the UAE, the broker would never have used us again, even though it was not our fault. In truth mechanical failures are very rare in our industry – most of the incidents and accidents are caused through human factors.

✈

On the ground in Liverpool before our Lucky Engine Failure flight!

VIP B757 – a large private jet

2 FLYING WITH DAD

Dad (on left) with 193 Sqn RAF Hawker Typhoon - 1944

I never flew with my father although I always wished I had done. That said I know of several pilots who have flown with a parent on a commercial crew, one of them whose mum was an airline senior training captain with an airline. Imagine how cool THAT would be?! These days of course, 'the age thing' kind of ceases to be an issue really, or rather I should say its importance has diminished. The main reason for this is the continuous use of modern CRM throughout the industry since the mid-'90s, which has flattened the much discussed 'authority gradient' in the cockpit.

The adverse effects of a steep authority gradient have been cited many times in case studies of airline incidents and accidents. Of course, experience still counts, even though we are all encouraged to attach less importance to age. There was a time though, as a young first officer when I used to observe a lot of my dad in the behaviour, mannerisms and leadership of the uniformed gentleman sitting in the left-hand seat...

✈

Operating a flight one day into Orlando, Florida, I listened to the latest ATIS weather broadcast from Orlando McCoy airport in Florida. Then I realised we had a real problem on our hands. There had been some severe storms and even now the wind was blowing hard. How hard? Well, the surface wind was still 'gusting' up to 40 knots across the runway. Bear in mind that in the early 1990s the airline Operations Manual stated a 'maximum demonstrated crosswind of 30 knots' but this was described as 'NOT limiting'.

Basically, the manufacturer's test pilots had managed to demonstrate landing the machine in wind conditions of 30 knots across. I had control of the sector inbound to Orlando, as Pilot Flying (PF), but when I passed the information over to the captain, he gently suggested that he should make the landing. I readily agreed, because frankly the thought of trying to put the aircraft down on the runway in that kind of wind seemed a big ask. I knew he was hugely experienced, in fact in some ways I thought of him as 'Dad' when we operated together. Only a couple of years later, I attended his '20,000 hours party' in Yorkshire – one of the finest aviators I ever knew.

We still had some time in the descent before we were to make our final approach to land, so we advised the cabin crew to secure the cabin early as it was going to be rough. The captain also made a very nice PA announcement to the passengers explaining that our landing in Orlando was not a certainty due to weather and we might have to Go-around – or possibly divert to Jacksonville in northern Florida. We had already copied the latest JAX weather report to confirm it was a good alternate for us. The crew secured the cabin and we started our initial approach towards the airport.

It was rock and roll all the way down – there were times when we saw 70 knots of wind on the instruments around 5,000 feet. Fortunately, the strong westerly reduced gradually as we descended further directed by a very calm ATC controller. His laconic voice informing us of the latest wind, including gusts at the runway, he also told us "two previous aircraft had Gone Around..." Meantime as we came onto final approach, I noted there was a huge amount of drift to remain steady on the inbound track to the runway. 'Dad' was doing a magnificent job of keeping the wings level and making the required adjustments to the pitch and power settings as we continued down the glideslope from 3,000 feet. Just nine miles to go now and he was flying manually as there was no way the autopilot could cope with this turbulence.

I looked across at him and thought *Wow, I've never seen you work this hard Tony! It looks like you're wrestling an alligator...*, which was quite appropriate as this was Florida after all. As Pilot Monitoring, I selected the flaps and gear as he called for them and kept up a continuous feedback of where the flightpath was headed, "Speed, Speed!" and "Glideslope one dot low... now One Dot HIGH!" then "...Centreline!" We were getting closer and the crosswind had dropped now to a steady 35 knots across in the final few hundred feet. Should we Go-around? Not yet, I reasoned, he was still flying well and we were essentially stable on final approach with gear down and Flap 30, even though I was viewing the runway almost through my side cockpit window.

came again, this time it was dark and menacing... *"Ahhh! The Force is strong with this one!"*

✈

At the end of the landing run, we received the taxi clearance and I vacated the runway slowly. After the previous dramas of the evening, there was no need to rush now.

"Captain, when did you start airline flying?" came the voice from my right.

"Well, I did my first Base Training on the B757 back in 1989" I looked across the cockpit smiling at my young colleague and continued "... First Officers used to reply, 'I was still at school' - but these days they say, 'I wasn't even born!'"

He calculated for a moment and nodded.

"I think... MY MUM was still at school back then!" To which I roared laughing. I had not heard that one before!

✈

BACKSTORY: I wrote this piece in November 2018. In fact, the Orlando flight took place back in 1991 flying for Air2000, the Marseille one in 2017. Everything I wrote happened just as described. Capt. Tony Storey and I had been teamed up together as a crew for our Air2000, B757 Type Rating in Singapore – Spring 1989.

I cannot tell you how much I learned from this man while operating as his First Officer. Always thought of him as 'Dad' and in fact there were similarities between Tony and my real father. Both of them had a wry sense of humour, great humility, and modesty.

Tony will cringe when he reads this, but he needn't worry. He is one of the best aviators I have ever flown with, he has the greatest respect from many of his contemporaries. He has now retired and lives comfortably in Yorkshire (God's own county) with Wendy his wonderful wife whom he refers to affectionately as... 'The Handbrake'.

✈

Captain Tony Storey Air2000 B757

3 POWERBACK AT AGRA

There was a complete dust storm outside the windows...

I wrote a story some time ago about the events of a diversion in India with a B757 while trying to carry out a charter flight into and out of Agra. This airport is the aerial gateway for tourists to visit the Taj Mahal – hence the title of the piece, "Monument to Love." Where the story ended was with our successful approach and landing into Agra, an Indian Air Force base. However, it was after landing that we confronted our biggest problem.

✈

Our taxi instructions from the tower after landing, led us down increasingly narrow taxiways, past rows of military hardware, towards the tiny civilian area of the airport. We were prepared for this as we had studied the airfield briefing, which stated there was only enough room on the tiny concrete civil ramp for one airliner and nothing else. What we were not

When I did this powerback, I really was on shaky ground, but I had considered every other way to get us out of there. Although undesirable, it was our last and only option. When we got back to Gatwick, I went to see the DFO and told him all about it. I also wrote up an Air Safety Report (ASR) describing the full circumstances of the event.

Note: Checkout YouTube to watch a powerback of a TUI B757 on the runway, after it overshot the exit for the apron on the Greek island of Skiathos. Captain I salute you. I don't know who you are, but you're a brave pilot.

https://www.youtube.com/watch?v=lqZ839Jc7DE&ab_channel=Cargospotter

Google Maps Satellite – positions of jets before powerback

4 YOU NEED TO BE BRAVE

Of course, in the old days only brave men and women flew aeroplanes. Back then it was dangerous... very dangerous. So naturally you would have to be brave (or stupid) to fly... But later it got safer, so you could be forgiven for thinking that maybe 'bravery' was not required. You would be wrong. But before we go there, let's think back - to the times when pilots like Louis Bleriot, Amelia Earhart and Charles Lindbergh strode the earth and plied the skies. They were giants, absolute GIANTS! Hard to imagine now, when you can purchase a ticket to fly across the surface of the planet at 500mph with ease.

In those days NOBODY could fly! Think about that. It was just something that was not available to 99.9 percent of the population of every country in the world. Regardless of how rich you were, the flying machines had not been invented, the research and development had not been done, it was just not possible. BUT... there were men and women who imagined the possibilities. And guess what...? Their dreams came true. They had the audacity to dream that the impossible, could be achieved... and they were right.

✈

Nowadays flying is such a routine activity it hardly bears mentioning. It is an ordinary way to travel, so commonplace. Just another form of transportation. And yet, and yet... it's still got something. In reality, when you start to learn how to pilot an aeroplane yourself, it is still very special. There is a kind of magic to it all. No, it's not a black art, but there is magic. It kind of takes your breath away. I remember distinctly, even now, that feeling... Fear, excitement, tingling, apprehension, worry and fear. Did I mention fear? It *was* frightening. In a Cessna 150 single-engined aircraft, we had got airborne from a grass strip, (Sherburn-in-Elmet) a trial lesson. Now the instructor said,

"Follow me through. YOU have control!" Well, I certainly didn't, but as I held on to the control wheel, it felt like - I was balancing on the head of a pin. Seriously, that was flying to me. No sense of speed, just balancing on the head of a pin! It was 1975 on a cloudy day in Yorkshire and there I was, at the controls of an aeroplane for the very first time – my head was spinning. I didn't dare move the control wheel, I just kept it horizontal and hoped and prayed we wouldn't topple off that pinhead.

Much later of course I entered the military flying training scheme. We knew what it was like to be afraid then, mostly of failure it must be said. In military terms, it's known as 'The Chop'. When you are on a military flying course the very last thing you want is for anyone to sense your fear. So, you learn to hide it, often with bravado. Declarations of confidence and over-the-top bullish behaviour are the norm. But all the time in your mind you are thinking, I can talk the talk, but can I walk the walk? With each successive achievement, each goal attained, and each course passed; real confidence in your own abilities builds. You get to the stage when you think, *'I could fly the box the aeroplane was delivered in!'* Later you encounter more challenges when you leave the relatively safe haven of the military and join the commercial arena. Now people look to you for your guidance and leadership as you progress from First Officer to Captain.

Like the time in Thailand while working for Air2000 in 1995 - when I realised that we had a problem with overflight permissions on our route from Bangkok, across Asia to Bahrain in the Gulf. The aircraft was on a turnround from Phuket, so passengers were waiting, the crew were ready and the handling agent was keen to get us away on schedule. It was night-time and we were a very long way from our home-base in the UK. The Dispatcher said he thought there might be a problem with the overflight permissions and asked me what I thought? A look at the paperwork showed one overflight permission listed, for entry into Pakistan Airspace that was several hours into the flight. I wanted to talk to our Operations centre in the UK, but this was in the days before mobile phones and the internet. The Dispatcher directed me to a payphone in the corridor near the airbridge.

The passengers were boarding our aircraft behind me as I started feeding coins into the machine. I distinctly recall standing there in my uniform, feeling ridiculous, trying to get a call through to Manchester from the Far East in the middle of the night. I had just got through to Ops; when the pips sounded and the line went dead, no more coins. Now what?! The aircraft was ready, refuelled with passengers onboard, pushback team connected, we had a valid flight plan all the way to the Gulf... I thought to myself, *take your courage in both hands son. Let's go!* A few minutes later, having received our ATC clearance and in as calm a voice as I could muster, I said to Nick the FO,

"Call for push and start please." As we climbed out of Bangkok, the night was beautifully clear. Mercifully we did not have bad weather to cope with also. ATC handed us over to Myanmar Control which informed us that we did not have overflight clearance for their airspace, and we should return to the Bangkok FIR. I was handling the radio as Pilot Monitoring while Nick did the flying. He looked across at me in the dark cockpit for guidance. With more confidence than I actually felt, I pointed at the next track across Burmese airspace and said,

"Keep going mate, we'll be fine!" On the radio I politely informed Myanmar Control that we did have valid permissions. They requested the company name and address. I spent much of the next seven hours reading out the company details to controllers at each FIR boundary, so they knew where to send the bills . Of course in many ways that is what overflight clearances are all about, countries collecting payment for using their airspace.

received our Air Operators Certificate five months earlier, so the company was in its infancy. I knew the Ops people were already pretty stressed out with a whole line of flying being disrupted by the diversion of our aircraft – so I was reluctant to call them now with this. It would not be easy for them to rustle up another big cab or a pair of taxis to come and find us in the back streets of Luton at that hour. Then suddenly, we were on the M1 and doing 70mph heading north towards East Midlands airport. Now, it seemed too late to do anything, apart from talking to him over his shoulder to keep him awake. I saw by his eyes in the mirror that he was falling asleep periodically. What a nightmare! These days it would be different.

Finally, we reached our destination and got out. There had been a couple of near misses enroute as the driver hit some fog patches at too high a speed and we all shouted,

"SLOW DOWN!" We were all glad to see the back of him and headed into the terminal. It was the usual pantomime getting through staff security with our overnight baggage at an airport where we didn't normally operate. Eventually the handling agent came to help us and escorted us Airside to the aircraft. Remotely parked away from the terminal, there it was, a Boeing 737-300 with EFIS, all shut down and closed-up. It looked like the previous crew had left in a hurry. We needed to arrange for replenishing of potable water, dumping the lavatory waste and cleaning the cabin which was a mess. Still we had some time in hand, because Ops reported the passengers had only just left Luton. I reckoned we had at least an hour and a half.

The printed flightplans arrived with the Dispatcher along with all the weather and Notams. The first officer and I could now decide on the fuel and complete our flight-planning for Geneva. In 1998, it was possible to get internet inside buildings, but not in your pocket. Although the weather there was fine, we ensured we had plenty of fuel loaded for the flight. At times like these when extraordinary circumstances prevail, we should not need to worry about 'minimum fuel' reserves too. In my experience, if one or two things start to go wrong, then there's much worse to come.

By the time the Dispatcher announced the passengers' arrival we had been ready for them for some time. Of course, it took ages for them to transit through the terminal as they had to be rescreened by check-in and security. I reassured the Cabin crew that they had my backing if any of the passengers became rude or abusive about the delay.

"Just do your best, keep smiling, welcome them onboard and remind them we had to come from Luton too, get them on our side!"

Very soon the cabin of this funny little start-up airline from Luton who painted the phone number on the side of the fuselage filled with skiers going out for the last couple of weeks of the season. They were not particularly pleased about trundling up the country to find their jet, but I did my best to assure them that safety was our top priority etc.

expecting was that there would be any other aircraft present when we parked up. Unfortunately, we spied a white bizjet parked on the apron that was going to restrict our manoeuvring considerably.

Following the marshaller's instructions, I used full lock on the 757 steering tiller to turn right through 90 degrees, prior to turning back left again the same amount. The first officer (and safety pilot) were both straining to look through the right-hand side windows to ensure it was all clear as we turned. As we turned fully to the left, I could see there was insufficient space to go very far forward as the corporate jet was blocking that side of the ramp. There was just enough room to get the nosewheels straight and then apply the parking brake. We shutdown the engines, switched off the seatbelts signs and completed the checklists. It was very quiet in the flightdeck as I said,

"Hmmm, that was fun. I wonder what time that bizjet is due to depart."

✈

We disembarked the passengers with suitable apologies for our late arrival and wished them well as we commenced the turnround. As I had flown into Agra, I briefed the new trainee First Officer that he was going to operate from Agra to Istanbul then left him in the cockpit to set up for the departure. I then went down the stairs to survey the parking situation. What I found made my skin go cold despite the hot temperature. I checked to see if they had a tug and towbar for a Boeing 757 – no chance, the nearest one was in Delhi. I looked carefully at the spacing on the ramp, there was not enough room for us to taxi left to avoid hitting the bizjet. It was only a few metres ahead of the nose of the 75'. It needed to be moved, no question about it. I found the dispatcher and asked him if the white jet would be moving soon... He replied affirmatively with a smile and a bobble of the head,

"Oh yes sir! In two days' time." I explained we could not wait for two days. He looked blank, so I asked to speak to the airport manager of the civil terminal. While the manager was called, I checked my watch, it was 10 am in Agra, but 5.30 am in London – there was no point in trying to call the duty management pilot at that time of the morning. Besides, what would he say? "Wait another two days..." I didn't think so.

Shortly after this, a smartly dressed man appeared and introduced himself as 'The Airport Manager' - I explained the problem and asked if we could somehow move the business jet, as we could not get past it without colliding with our right-hand engine.

"But Captain you can turn very tightly to the left... no?"

I paced it out and showed him where our right engine would track over the ground, even with full steering lock on the tiller. It would contact the left wing of the jet, no doubt about it. The obstacle had to move. He looked very unhappy and said that the little white jet belonged to some government bigwig and was all locked up for the next two days. Knowing he (we) had 230 grumpy, delayed passengers in his terminal, I racked my brains for a solution. Yes, there was one, but I had never attempted it before.

I went to the rear of the maingear tyres to determine how far...

"A powerback is THE ONLY WAY we are going to get out of here," I explained it to him, "But I NEED your permission to do it. I warn you now my friend, it will be VERY NOISY!" He agreed.

My head was swimming as I walked away. *Are we really going to do this?*

Powerbacks' with jet airliners went out in the 1980s. In 1989 when I completed my initial 757 type training, the manoeuvre was still included in the Boeing Flight Crew Training Manual. However certainly, by this time in 2006, any mention of 'powerback' had been deleted. The Safety Pilot and 'flying spanner' (travelling engineer) agreed with me that there was insufficient room to taxi out without collision, making a powerback the only solution. So, I got them and a couple of the cabin crew down onto the ramp and briefed them on the next step.

"Look guys, we HAVE to clear the ramp of every little stone, rock, pebble and chip – it MUST BE as clean as possible!" So, we carried out an intensive FOD Plod of the entire concrete area that we would be using for the powerback, especially along the path the engines would track. I noted with dismay however that there was only so much we could do. In India dust settles on any piece of flat ground. Dust coated the apron, but there was no time for that now. At least we collected and dumped all the little stones and pieces of debris. The ramp was as clean as we could make it in the time available.

After that I went to the rear of the maingear tyres to determine how far we could reverse without the rear wheels going off the edge of the concrete, - a mere 15 paces, 20 maximum. While the crew were prepping the aircraft to fly, I was consumed with worry about whether I was doing the right thing. But what else could I do? The poor passengers had been delayed for 24 hours and to now send them back to their hotels would likely provoke a riot. Plus, the company would lose a jet from the programme for two days... Just not acceptable – a powerback would solve all our problems.

'*It's just that I've never done one before*', came the little voice in my head. '*Stop it! Get on with it!*' I scolded myself. '*Let's get this thing done!*' Now I knew it was vital to know where to stop the aircraft reversing, but we had nil rearward vision. And of course I could not trust the marshallers. They had put us in this mess in the first place.

So, I walked over to the far right-hand side of the concrete. Not very far in reality as it was such a small area. I placed a large lump of spare concrete on the edge of the ramp directly abeam the flightdeck windows, after which I marched 15 paces backwards in the direction we would be reversing to plant another lump. One last look over at the space behind the maingear bogies... *Hmmm... doesn't look like a lot of room to me...* then back onboard the jet, just as the passengers started arriving. Very soon the cabin was full, the refuelling had been done and cargo holds closed, just the main L2 door remained open.

I addressed the passengers from the forward galley in full view of them all, not something that I enjoyed, but a necessary evil. They have a right to see us face to face and it helps to get them to empathise with the crew. As I was talking to them, having carefully planned the wording of the announcement, I considered whether to mention the powerback. I decided against this because it would only be loud for a few seconds after engine start and maybe they wouldn't notice. I did however brief the cabin crew to expect it to be plenty noisy after start up. They were already used to moving backwards after start, so that should be normal for them, *provided of course I didn't sit the damned thing on its tail.*

Now all four of us were in the flightdeck, strapped in and the L2 door closed. With the Cabin Secure, I looked outside where the groundcrew signalled we were clear to start engines. It was unusually quiet as the RB211s spooled up and once idling, we completed the after-start procedures, *here we go...* I waved off the groundcrew who took the chocks, standing well clear over to our left. During my briefing to the guys in the cockpit I re-emphasised that I MUST have my feet flat on the floor during the powerback and they agreed to check me. This would prevent me from using the brakes to arrest our rearward travel, which could tip the whole airliner onto its tail! The feeling was completely weird however as I released the parking brake, put my feet flat on the deck and engaged reverse thrust on both engines. No time for doubt now. *'Let's get this thing done!'* I said to myself. Outwardly I may have seemed confident, but inside I was trembling and quivering with anxiety, *'My God what a mess! Let's pray this works... never done this before'*. I gradually applied reverse thrust; the noise level increased to a muted roar behind as the engines strained to pull 110 tonnes of airliner backwards.

"It's not moving!" I stated in as matter of fact a voice as possible. Four pairs of eyes looked to the right at the concrete block abeam our position. The dust clouds being blown forwards were intensifying, but we could still see the rock. "We need more power..." and I pulled the reversers up higher. *'My God she's heavy, c'mon baby, you can do it'*. There was a complete dust storm outside the windows and the roaring behind us was incredible. Huge powerful vibrations could be felt through the whole ship.

"NOW! She moves...!" Gradually rolling backwards, we all looked for the other rock and slowly it appeared. I reduced the power as soon as we overcame the inertia. I glanced out front and the little white jet had completely disappeared in a brownout. As the flightdeck came abeam the second rock, I cancelled reverse and engaged forward thrust. The temptation to use my feet was intense. There was a momentary doubt that we would not stop going backwards in time, but I should not have worried, the aircraft was really heavy and stopped easily. Now as we focused on the forward motion and prepared to taxi, we saw one very dirty private jet! Totally covered in dust, the sight did not make me feel at all comfortable. Thank God we had picked up all the stones. I started to increase the power and as we felt the machine roll forward, I wound on full steering lock to the left and followed the marshaller's instructions to exit the tiny apron. We collectively breathed a huge sigh of relief as we straightened up on the taxiway. Next stop, Istanbul Ataturk airport... we had escaped!

✈

BACKSTORY: This event was in 2006, we were flying for Astraeus Airlines out of Gatwick. I wrote the story in 2019. It goes without saying that reversing a Boeing 757 is not only unusual, but hazardous.

I remembered the older pilots discussing it in 1989 when it was still included in the Boeing Flight Crew Training Manual. Some Airlines used to powerback off the gate in the USA in the '80s, although that was mostly with DC9s and the like.

When I did this powerback, I really was on shaky ground, but I had considered every other way to get us out of there. Although undesirable, it was our last and only option. When we got back to Gatwick, I went to see the DFO and told him all about it. I also wrote up an Air Safety Report (ASR) describing the full circumstances of the event.

Note: Checkout YouTube to watch a powerback of a TUI B757 on the runway, after it overshot the exit for the apron on the Greek island of Skiathos. Captain I salute you. I don't know who you are, but you're a brave pilot.

https://www.youtube.com/watch?v=lqZ83gJc7DE&ab_channel=Cargospotter

Google Maps Satellite – positions of jets before powerback

4 YOU NEED TO BE BRAVE

Of course, in the old days only brave men and women flew aeroplanes. Back then it was dangerous... very dangerous. So naturally you would have to be brave (or stupid) to fly... But later it got safer, so you could be forgiven for thinking that maybe 'bravery' was not required. You would be wrong. But before we go there, let's think back - to the times when pilots like Louis Bleriot, Amelia Earhart and Charles Lindbergh strode the earth and plied the skies. They were giants, absolute GIANTS! Hard to imagine now, when you can purchase a ticket to fly across the surface of the planet at 500mph with ease.

In those days NOBODY could fly! Think about that. It was just something that was not available to 99.9 percent of the population of every country in the world. Regardless of how rich you were, the flying machines had not been invented, the research and development had not been done, it was just not possible. BUT... there were men and women who imagined the possibilities. And guess what...? Their dreams came true. They had the audacity to dream that the impossible, could be achieved... and they were right.

✈

Nowadays flying is such a routine activity it hardly bears mentioning. It is an ordinary way to travel, so commonplace. Just another form of transportation. And yet, and yet... it's still got something. In reality, when you start to learn how to pilot an aeroplane yourself, it is still very special. There is a kind of magic to it all. No, it's not a black art, but there is magic. It kind of takes your breath away. I remember distinctly, even now, that feeling... Fear, excitement, tingling, apprehension, worry and fear. Did I mention fear? It *was* frightening. In a Cessna 150 single-engined aircraft, we had got airborne from a grass strip, (Sherburn-in-Elmet) a trial lesson. Now the instructor said,

"Follow me through. YOU have control!" Well, I certainly didn't, but as I held on to the control wheel, it felt like - I was balancing on the head of a pin. Seriously, that was flying to me. No sense of speed, just balancing on the head of a pin! It was 1975 on a cloudy day in Yorkshire and there I was, at the controls of an aeroplane for the very first time – my head was spinning. I didn't dare move the control wheel, I just kept it horizontal and hoped and prayed we wouldn't topple off that pinhead.

Much later of course I entered the military flying training scheme. We knew what it was like to be afraid then, mostly of failure it must be said. In military terms, it's known as 'The Chop'. When you are on a military flying course the very last thing you want is for anyone to sense your fear. So, you learn to hide it, often with bravado. Declarations of confidence and over-the-top bullish behaviour are the norm. But all the time in your mind you are thinking, I can talk the talk, but can I walk the walk? With each successive achievement, each goal attained, and each course passed; real confidence in your own abilities builds. You get to the stage when you think, *'I could fly the box the aeroplane was delivered in!'* Later you encounter more challenges when you leave the relatively safe haven of the military and join the commercial arena. Now people look to you for your guidance and leadership as you progress from First Officer to Captain.

Like the time in Thailand while working for Air2000 in 1995 - when I realised that we had a problem with overflight permissions on our route from Bangkok, across Asia to Bahrain in the Gulf. The aircraft was on a turnround from Phuket, so passengers were waiting, the crew were ready and the handling agent was keen to get us away on schedule. It was night-time and we were a very long way from our home-base in the UK. The Dispatcher said he thought there might be a problem with the overflight permissions and asked me what I thought? A look at the paperwork showed one overflight permission listed, for entry into Pakistan Airspace that was several hours into the flight. I wanted to talk to our Operations centre in the UK, but this was in the days before mobile phones and the internet. The Dispatcher directed me to a payphone in the corridor near the airbridge.

✈

The passengers were boarding our aircraft behind me as I started feeding coins into the machine. I distinctly recall standing there in my uniform, feeling ridiculous, trying to get a call through to Manchester from the Far East in the middle of the night. I had just got through to Ops; when the pips sounded and the line went dead, no more coins. Now what?! The aircraft was ready, refuelled with passengers onboard, pushback team connected, we had a valid flight plan all the way to the Gulf... I thought to myself, *take your courage in both hands son. Let's go!* A few minutes later, having received our ATC clearance and in as calm a voice as I could muster, I said to Nick the FO,

"Call for push and start please." As we climbed out of Bangkok, the night was beautifully clear. Mercifully we did not have bad weather to cope with also. ATC handed us over to Myanmar Control which informed us that we did not have overflight clearance for their airspace, and we should return to the Bangkok FIR. I was handling the radio as Pilot Monitoring while Nick did the flying. He looked across at me in the dark cockpit for guidance. With more confidence than I actually felt, I pointed at the next track across Burmese airspace and said,

"Keep going mate, we'll be fine!" On the radio I politely informed Myanmar Control that we did have valid permissions. They requested the company name and address. I spent much of the next seven hours reading out the company details to controllers at each FIR boundary, so they knew where to send the bills . Of course in many ways that is what overflight clearances are all about, countries collecting payment for using their airspace.

We had the flightdeck full of airways charts like never before as we searched for the next VHF frequencies along the route. I noticed my hand was trembling in the darkness as I hit the transmit button to broadcast our intentions. It was slightly more relaxing at cruise altitude as we knew that we would be very hard to stop now as we tracked along the airways at 500+mph. Our confidence was building as we approached Pakistan, our penultimate ATC region before we arrived into the Arabian Gulf. As before, we loudly and clearly transmitted our data, but this time we had a real overflight clearance number to read out. After reciting it with the correct phonetic alphabet, I waited for their reply, which was,

"That clearance number was valid two weeks ago and was for one flight only!" Aaargh! My heart skipped a beat, but after all we had been through, we were not stopping now. As on previous occasions I apologised profusely, read out the company details and we continued.

On handover to the Gulf ATC and with descent clearance into Bahrain, I knew we had made it. Hearing an American radar controller was like coming home. We touched down just as the sun was rising in the east behind us. It was a real 'high-five' moment for me and Nick. We had just flown over 3000 nautical miles without any overflight clearances whatsoever! The Diplomat Hotel in Bahrain was a very welcome sight as our crewbus pulled up outside.

PS. Three days later I was in the chief pilot's office explaining what we had done and why. Fortunately, he viewed it as we did and was very supportive. Feedback was given to the Flight Ops Department to ensure there was no repeat.

✈

BACKSTORY: Nick the FO on this flight became a senior flight operations manager for a UK airline later in his career. This night flight against the odds happened in Air2000 not long before I resigned and went to Alitalia on the B767 Nordstress contract.

In Chief Pilot Bob Screen's office, I recounted the story for him and when I said I'd been worried about being intercepted by Pakistani Air Force fighters at 37,000 feet, he was dismissive and laughed, "They haven't got anything that could climb that high!" I doubt this type of flight could ever occur today. It was a moment in time.

✈

5 APRIL FOOL'S DAY

You know there are some days when you really wish you had stayed at home. I am not superstitious, although I know some that work in aviation who are. You will notice that on nearly every airliner the row numbering system goes from 12 to 14 without a break – there is no row 13. This is one less issue for the ticketing system to resolve.

1st April 1998 was a day to remember from many angles, not least of which was the weather. We were due to operate a three-sector day from Luton with an early departure. As I drove carefully to the airport through the fog, I was already thinking about 'Low Visibility Procedures' (LVPs). When we reported for duty it became apparent that our day would be a long one, although I could not have predicted just how extended our duty would become. The crew were naturally looking forward to our Barcelona night-stop at the end of the third sector, we just had to fly the Geneva and back first. Some chance of that, as I called Ops to find out where the aircraft was.

"Just leaving the Luton holding pattern, diverting to East Midlands Airport. The aircraft is *CAT 1 ILS only and cannot do a Low Vis approach..." Was the reply from Ops.

Well this was a shock, however they already had a plan.

"Your crew will go in a taxi to East Mids. and we will bus the passengers there too. By the time you go to Geneva and back, the fog at Luton is forecast to clear so you can fly a CAT 1 approach..." It sounded logical enough, although trucking everyone up the M1 in the fog would take time. Our transport of delight outside the terminal, turned out to be a disappointment. An ancient minicab, of indeterminate Japanese origin. After five of us crammed inside with our overnight bags, it was packed to the roof. The front seat next to the driver was piled high with baggage that could not be stowed in the back. As if that was not bad enough, we were overpowered by the smell of stale cigarette smoke and body odour from the unshaven driver. Even worse was the definite scent of alcohol on his breath.

✈

Well, now I had a problem; what to do? Clearly, he had been called from home very early, bear in mind our report time was 0530 for an 0615 departure. I guessed it was the lingering smell from the night before, rather than that he'd been drinking this morning. Although his driving seemed okay at first, he was prone to wandering from lane to lane as we travelled through the back streets of Luton heading for the motorway. Hindsight is 20/20 vision of course and I wish I had called a halt to it then. I should have instructed him to pull over to the side of the road so we could take another taxi.

The trouble was, we were already late. The airline had just six aircraft in those days and disruption to the programme was always costly. We had only received our Air Operators Certificate five months earlier, so the company was in its infancy. I knew the Ops people were already pretty stressed out with a whole line of flying being disrupted by the diversion of our aircraft – so I was reluctant to call them now with this. It would not be easy for them to rustle up another big cab or a pair of taxis to come and find us in the back streets of Luton at that hour. Then suddenly, we were on the M1 and doing 70mph heading north towards East Midlands airport. Now, it seemed too late to do anything, apart from talking to him over his shoulder to keep him awake. I saw by his eyes in the mirror that he was falling asleep periodically. What a nightmare! These days it would be different.

Finally, we reached our destination and got out. There had been a couple of near misses enroute as the driver hit some fog patches at too high a speed and we all shouted,

"SLOW DOWN!" We were all glad to see the back of him and headed into the terminal. It was the usual pantomime getting through staff security with our overnight baggage at an airport where we didn't normally operate. Eventually the handling agent came to help us and escorted us Airside to the aircraft. Remotely parked away from the terminal, there it was, a Boeing 737-300 with EFIS, all shut down and closed-up. It looked like the previous crew had left in a hurry. We needed to arrange for replenishing of potable water, dumping the lavatory waste and cleaning the cabin which was a mess. Still we had some time in hand, because Ops reported the passengers had only just left Luton. I reckoned we had at least an hour and a half.

The printed flightplans arrived with the Dispatcher along with all the weather and Notams. The first officer and I could now decide on the fuel and complete our flight-planning for Geneva. In 1998, it was possible to get internet inside buildings, but not in your pocket. Although the weather there was fine, we ensured we had plenty of fuel loaded for the flight. At times like these when extraordinary circumstances prevail, we should not need to worry about 'minimum fuel' reserves too. In my experience, if one or two things start to go wrong, then there's much worse to come.

By the time the Dispatcher announced the passengers' arrival we had been ready for them for some time. Of course, it took ages for them to transit through the terminal as they had to be rescreened by check-in and security. I reassured the Cabin crew that they had my backing if any of the passengers became rude or abusive about the delay.

"Just do your best, keep smiling, welcome them onboard and remind them we had to come from Luton too, get them on our side!"

Very soon the cabin of this funny little start-up airline from Luton who painted the phone number on the side of the fuselage filled with skiers going out for the last couple of weeks of the season. They were not particularly pleased about trundling up the country to find their jet, but I did my best to assure them that safety was our top priority etc.

"Ladies and gentlemen, welcome onboard. The good news is that the weather in Geneva is lovely and we will be on our way soon, thank you so much for your patience this morning..."

With the passengers all onboard and the aircraft holds closed, all we needed now was the loadsheet. As the Dispatcher came up the airstairs I was hoping he was here to deliver it. No such luck. He came in the cockpit,

"Skipper... *You've got a fuel leak!"* I felt sick. With all eyes on me through the flightdeck door, I refrained from screaming what I was thinking. I simply took a deep breath and said,

"Really...? That WOULD be unfortunate... I will take a look" As I exited the flightdeck I felt 150 pairs of eyes boring into me as I promptly turned right and went down the steps. I was thinking, *'Please God Noooo!'* And they were all thinking, *'Now what?'* and I didn't blame them. *Let's hope this really is an April Fool's joke* I mused as I went straight to the left-hand wing to look. Surely, it's just some dripping condensation...

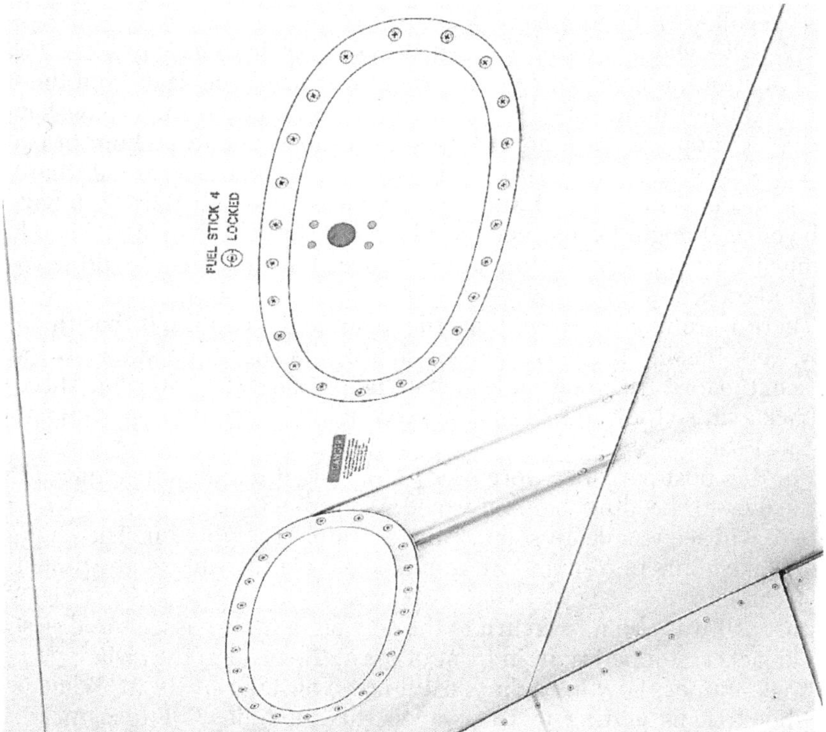

Boeing 737 underwing fuel panels

I arrived under the wing and looked up. Nope! It really is. *'It's a bloody fuel leak! Seriously??! Aarrghh!'* Again, the voice in my head screamed aloud, but I had to APPEAR LIKE I was in control. They all look to the Aircraft Commander, 'The Skipper', for leadership. I followed the line of fuel leaking from the under-wing inspection panel tank seal, all the way to the side of the left engine and it was dripping on to the concrete apron forming a puddle. We hadn't seen it on our pre-flight walkround because the fuelling had not been completed. The Dispatcher found it during his final pre-flight check after the tank had been filled up. I turned to him and said,

"Please call the engineers to come and have a look as soon as possible – I will explain to the passengers." The leak was more than just a drip and I could see that to stop it leaking, they would have to partly defuel the tank, unscrew the access panel (lots of little bolts) and then change the soft seal for a new one IF it was available... Minimum couple of hours work – damn.

At times like these it is always best to deliver bad news face to face with the customers. But that still didn't make it any easier as I picked up the forward galley PA hand mike.

"Well everybody, I wish this really WAS an April Fool's day joke... but it's real I'm afraid. I have just been informed we have a fuel leak from the wing tank on the left-hand side and the engineers are on their way to check it out. Whatever happens, it's going to take some time, probably an hour or two, so we will ask you to disembark again into the lounge at the terminal where you will be more comfortable. I am so sorry. Please take all your cabin baggage with you as it might be another aeroplane you come back to later. Thank you so much for your cooperation and patience. The groundstaff will keep you updated on progress."

There were moans throughout the cabin and I could tell that the cabin crew were tensing themselves to be the butt of the complaints, so I stood there in the forward galley with them to help deflect the criticism. There was no back exit, so they all had to come through the forward door down to the waiting buses.

For the most part they were pretty stoic about it and quite a few of them thanked us for keeping them informed. We replied,

"We will see you again soon..." But privately I doubted it. The engineers would take hours to fix it and we would be out of duty time by the time it was serviceable.

Just as I was about to return to the cockpit, a male passenger, dressed in his ski jacket, stopped right in front of me at the head of the aisle.

"Well James, how long do you think it will take to fix it? What is the likelihood of us getting to Geneva on this jet today?" I recognised him instantly as my mind reeled... *Blow me down with a feather! That's all I need. Captain Ian Gibbs our airline's CAA Flight Operations Inspector... HE is one of our passengers today. Oh No!*

✈

Honestly, you could not have made it up. WHAT were the chances of that?! I was frank with him as I kept my voice low. I told him I didn't think we would be going to Geneva on this aircraft today, to which he replied,

"Thank you for that James, I appreciate your honesty. We will make other plans. But I'm REALLY looking forward to seeing the postflight paperwork on THIS ONE!" He left the aircraft chuckling to himself as I stood there shaking my head. *No way!* But it was true. Our CAA Inspector had just been on board our aircraft which was stranded with 'a Tech problem', after a weather diversion, with a crew that was running out of hours... Can this day get any worse? There would be no chance of sharp-pencilling anything here!

Soon after the cabin was empty, the engineers arrived. As I expected; there was plenty of sucking of teeth and shaking of heads. They didn't know if there were any spare parts available, where they would get them, how long it would take or even how long to defuel the tank. The crew decamped to the terminal where we blagged our way into the British Midland Crewroom. We happily plotted up in a corner with all our baggage and informed Luton Ops where to find us. With a coffee machine, comfy seats, and snack bar close by, we could veg-out. It was clear from looking at the duty timings that we would be out of hours before too long.

"Maybe we will fly empty back to Luton?" The cabin crew's imagination was working overtime too. Imagine our disappointment when Luton Ops called only two hours later. The voice on the phone sounded pleased,

"They've fixed it, you're good to go!" Oh damn! We had all wound ourselves down and were convinced we wouldn't be flying 'Yankee Broken'**. The cabin crew dreaded the idea of facing 150 grumpy ski-holiday pax. We agreed that the ski-crowd are among the most demanding. These frequent fliers know what they want and what to expect in the way of service standards. They also do not hesitate to complain if they don't get it. I gave the crew a pep-talk.

"It's only 90 minutes to Geneva and then they will be off the aircraft, you will never see them again. Just keep smiling and try to be as nice as possible. They will probably be exhausted and just be pleased they're travelling today after all. And remember, we are in this together. It is still US! We are the crew who have lost our nightstop in Barcelona!"

In the cockpit we got in contact with ATC and refiled the flight-plan, checked there was no slot and started the pre-flight sequence again. The Techlog was signed off by the engineers. Fair play to them, they had done a magnificent job against the odds – no wonder the guys in Luton were pleased. If nothing else went wrong, we were just in duty hours to get this rotation done, two sectors, back into Luton.

We crosschecked the Luton latest forecast and it was good enough to give us a CAT 1 approach when we arrived back from Geneva. Let's go! As we had some time, I made sure to stand with the cabin crew in the forward galley during passenger boarding, I didn't want them to face it alone. Also, if there were any obnoxious travellers I would be able to see them immediately and get them offloaded – no way would our crew be abused today. In fact, the passengers were lovely, they knew we were the same crew and were surprised to see us. As I went back into the flightdeck I started to feel better about things. I shouldn't have let my guard drop.

6 TIME TO SPARE

I don't normally fall asleep during the pushback, but this time it was different. I was sitting in Row 20 of a Lufthansa flight which was intending to fly from Athens to Frankfurt where I would change to go on to the USA. The pushback was 35 minutes late and during that time I had nodded off. There was now a protracted interval after the engines were started and I feared the worst -- the crew were struggling with a technical problem. Sure enough a few minutes later, over the PA came the hesitant tones from the flightdeck... *"Meine Damen und Herren - hier spricht Ihr Kapitän..."* I would have to wait for the English translation as I don't speak German, but already I knew this was not a good sign...

✈

I have had my fair share of delayed flights while operating as crew over the years. Not all of them due to technical reasons of course. The one I prefer the most is when the weather prevents our departure. This usually takes the form of some dramatic cloudburst, often overhead the airport making it senseless to fly. The other advantage to this is that the passengers know why we are not going anywhere, so there are no awkward explanations to make.

If they had not yet boarded we would normally leave them in the terminal where they would be more comfortable. It's also easier on the crew and I encourage them to put the kettle on and relax. Meanwhile the aircraft doors are closed to prevent a deluge of water entering the cabin. On a recent turnround in Italy, we were late in closing the front door after our last inbound passenger had left the plane. There was a mini Tsunami which threatened to enter the flightdeck. There was a flurry of activity as we built a paper towel dam consisting to stop the water coming in. The big danger here of course is that free liquid in the cockpit can drain through the floor into the dreaded E&E Bay and cause electrical short circuits etc.

Delaying the boarding of the passengers while a storm passes through the airport has other advantages also. Passengers may decide not to travel due to anxiety about the weather or because they would miss their connections for onward flights. If they are still in the lounge, then it makes the Dispatcher's job easier to find hold baggage and remove them and the baggage from the load figures. We cannot carry unaccompanied baggage without specific extra security screening. In addition, if they have not entered the cabin, then extra cabin security sweeping of the area is not necessary. Yes, on the whole, this is the one I like the best.

A trickier situation comes when the weather at destination is an issue. This should be apparent prior to departure and then maybe the takeoff is delayed waiting for the weather to clear. Now comes the difficult part – explaining the delay to the passengers and keeping them happy. Usually they have boarded the aircraft by this point, which means in their minds they are 'travelling' and thinking... "Let's go!" At an early stage it is necessary to get a grip on the situation to manage their expectations. The captain should inform the cabin occupants of the reason why they are waiting and then keep them updated regularly on conditions at destination. All efforts should be made in coordination with the cabin crew to ensure the pax are kept comfortable, permitted to use the lavatories, mobile devices etc. Special attention should be given to keeping the cabin temperature cool enough to prevent people overheating.

Note that the PA announcements should be made with empathy such that the passengers understand that the crew want to fly to destination just as much as they do. This will prevent them from attributing blame and giving the cabin crew a hard time. After the PA from Flightdeck, the cabin crew should repeat it word for word in the local language to ensure that there are no misunderstandings. In practice it is possible to keep the situation calm and well controlled for quite a long period of time if necessary. The key to the problem is information. If they are informed and comfortable, the passengers will be on your side.

From the operational perspective though, ground delays are hard to bear. When the airframe is stationary on the ramp, ready to fly, you are losing time at the rate of seven miles per minute – time you can never recover. Consider that the economical cruise speed of a modern airliner is approximately 94 percent of the maximum speed at high altitude. To increase speed by another Mach 0.03 costs an excessive amount of fuel and in the course of a flight of say two hours would only save a few minutes – hence there is no point. That being said, from the PR angle it is often good to tell the passengers the crew are doing all they can to minimise the delay for arrival, which is true of course. Pilots can employ direct routings along the airways (cutting corners with ATC consent) and expeditious arrival vectors; but cruising faster just wastes the planet's resources.

I learned much about the best ways to handle ground delays from my five years spent in the right-hand seat of the B757 with a leading Charter Airline in the UK. Valuable time indeed. I saw my very experienced flightdeck colleagues manage many delayed flights with professionalism and diplomacy. In those challenging Summers back in the 1990s, protracted ATC slot delays were much more common than they are now. The air traffic system across Europe is noticeably more efficient and that is despite a much-increased volume of traffic through all the airports. When the handling agent informs us that we have a slot which is perhaps 30 minutes after scheduled time of departure, there is no question about it. We simply board the passengers and call ATC asking them to send a 'Ready Message'. In 75 percent of cases the slot improves or disappears completely – a far cry from the bad old days.

✈

Going back to my Lufthansa experience – the Captain announced "This Aircraft is AOG and we cannot fly today. Safety is our first priority..." etc.

Although the term AOG* may not have meant much to other passengers, it was fine for me and assisted in completion of my compensation claim for €600 in accordance with European Regulation EC 261/2014. It simply confirmed that the reason for the delay was technical, although the skipper would have been better advised to use terms more understandable in his PA to the customers.

Thinking back to the '90s reminds me of a delay I experienced flying with Capt. Peter Walker. With the passengers all aboard, he and I were sitting there ready to pushback for the best part of an hour, enjoying a cup of tea when he calmly looked across to me, smiled and said,

"You know what they say. Got time to spare? Go by air!"

✈

*AOG – Aircraft On Ground, due to technical malfunction requiring rectification before flight.

BACKSTORY: My failed flight was actually operated by Sun Express on behalf of Lufthansa. To be fair, Lufty rebooked me on another Star Alliance carrier to fly the next day. They put me up in a hotel and gave me meals, plus reimbursed my full fare for the outbound ticket. I was happy about all that. I did miss out on a day of my vacation though and as a friend once said to me, "Every day you spend waiting in a hotel... that day will never come again in your life you know."

✈

Favourite runway for takeoff? Palermo RW07 – scenery, stunning!

7 DIFFERENT STANDARDS OF AVIATORS

There has been a great deal written about the Boeing 737MAX in recent months in the aftermath of the two 'hull loss' accidents and resultant tragic loss of hundreds of lives. I will not speculate on the cause/s of the accidents. It is enough to say that it would be unique for there to be only one cause. In every aviation accident, there are statistically between 7 and 11 links along the chain that leads to the final tragic event.

Until post-crash investigations are complete we shall not talk specifics. There was however, a public statement by the manufacturer recently that gave me food for thought. The Boeing company implied that 'Standards of Aircrew Training and Line Operation' were of a lower quality in airlines of developing nations. Basically, they seem to be stating that not all airlines pilots are equal – in short, airline pilots were not as competent in some parts of the world as in others. The initial reaction of intelligent observers would be to regard this statement with scepticism as being politically motivated, but that is too simplistic.

Of course, the training and standards issue is complicated when considering what happens in different countries. Common sense tells us it is essential to compare apples with apples. Objective assessment of competence/performance by pilots cannot be made in isolation, the influence of cultural factors has a bearing in addition to the local theatre of operations.

All that being said, I have been lucky enough to have acquired unique insight into training and checking standards of many airlines across the globe. It has been my privilege to work with airline crews of countries from the Far East, Middle East, Africa, Australia, former Eastern Europe, the Americas (North and South), plus of course airline companies that operate under the remit of EASA. My contact was not always in that of a training/checking role, but I did not fail to observe differences in standards when they were present. The one common factor throughout was the Boeing. I have operated with and taught/checked aviators in B737 (classic EFIS & NG), B757 and B767 short and long-haul variants.

After thousands of hours in Boeing simulators and flightdecks with my colleagues from so many other nations I speak from experience. I can honestly say that, yes, there are different levels of competence and performance evident in different states. And it is the case that sometimes the standard of operating is much lower than that which applies in EASA-land. Not only in the arena of technical expertise, but also just as importantly in Crew Resource Management (or Cockpit Resource Management if you prefer) – otherwise known to all as 'CRM'.

You may recall the Asiana Airlines B777 accident at San Francisco, when a perfectly serviceable aircraft operating in near-perfect weather conditions was flown into the sea-wall short of the runway threshold. The damning verdict of the NTSB accident report was that mismanagement of the approach and inadequate monitoring of airspeed led to the crash of Asiana flight 214. It would be a dream come true to say that all pilots and airlines from every state are operating to the same standard as those from Europe. I would be deluding myself to make that assertion. I live in the real world. There are so many factors at work – training standards, operational oversight and culture (both national and corporate) to name but a few. Where there are problems, they can be long-standing and deeply ingrained: such issues are not likely to be fixed overnight – even when there is a genuine willingness to accept that there is a problem at all.

There is a familiar pyramid graphic which comes to mind – known as Heinrich's Law. It states that for every accident (at the top of the pyramid) there are a much greater number of lesser incidents and even unreported incidents lower down. Experienced aviators will have seen it and it is used to visually display the numerical ratio of commercial airline 'accidents' compared to 'events'.

The number of accidents at the top of the pyramid is incredibly small of course, which supports the notion that airline flying is statistically the safest form of transport on the planet. What I'm getting at here, is this. There have only been 393 B737MAX airframes operating with the airlines thus far. During all those thousands of sectors in all weathers, day and night, by regular line pilots, by Heinrich's Law there must have been a significant number of reportable incidents which *MIGHT* have led to Loss Of Control Inflight (LOCI). The airline companies and the regulator (FAA) know the figures. In fact, so does the manufacturer of course. Hopefully the investigations into the two fatal B737MAX accidents will make the data on any such incidents public and decide whether they had any relevance to the accidents. Was there a difference in how differently-trained crews reacted to similar incidents?

On the positive side, ultimately, I have confidence in the integrity of the aviation accident investigation system and of those individuals working within it to define the genuine causes and contributory factors. They will also recommend a strategy to prevent future occurrences. Again, without speculating on causation, it is possible that *if* pilot training standards are deemed to be a factor in these two accidents (and it can only be an 'if' at this stage) – then appropriate remedial actions will be recommended for the future*.

And if such recommendations are made, they must be acted upon – regardless of politics, culture, cost, or any other non-safety related consideration. One of the worst features of the recent losses is that for so long our industry has been blessed with successive improvements in all areas technical. Each successive product of the airframers has come with a raft of better design features which makes the operation of the machines safer and more efficient.

Indeed, one of the major reasons for the emphasis upon modern CRM comes from the fact that so few of the major accidents of the past 30+ years have been caused by engineering/mechanical failures. Now Pilot/Crew Error and miscommunication/s within the team are cited as the largest single causation group. We are blessed in the UK with having some of the best qualified people in the industry with whom to work. Not only that, but UK companies and regulator have a good track record of putting into practice lessons learned from previous accidents/incidents. We should take nothing for granted, however when fast growing airline companies from around the world send their raw recruits to UK companies for training, that must say something positive about how the UK's training standards are perceived worldwide.

In all things aviation, reality is important. In every scenario of our profession, we always try to establish the facts before drawing conclusions. I leave you one fact. The last recorded accidental death onboard a UK-registered airliner was on 8th January 1989, at Kegworth near East Midlands Airport. That was over 30 years ago. I can't think of any other nation that comes close.

*Boeing has commenced an extensive support programme for the reintroduction of the B737MAX back into service. Currently they are sending hundreds of experienced 'Boeing Instructor Pilots' to customer airlines all over the world. There is no doubt that this will make a positive difference.

BACKSTORY: I penned this opinion piece in May 2019. We were all shocked at what occurred with the 737MAX and deeply saddened for the victims and their families. We Boeing loyalists who had flown their aircraft with utmost confidence for so many years, were devastated to learn of the company's apparent manufacturing and design misjudgements.

8 BLESS THE WEATHER

It is what makes our job interesting!

I keep hearing the words, “It’s been a funny summer…” usually by friends of mine in the UK. The last time there was such a hot, dry summer in Britain was 1976. I know, I was there - it was hot! This year has been exceptionally warm and dry, but not in all areas of Europe. By contrast in the Southern part of the Mediterranean there have been dramatic storms and days of continuously unsettled weather. Strong surface winds have occurred when they were not supposed to, bouts of heavy rain in the middle of summer in Italy and Greece – unheard of. Naturally everything points towards the much-debated Climate Change, or ‘Global Warming’ as it used to be called. As has been asserted many times, the “Warming” part of the phenomenon results in more extreme weather patterns, rather than raised temperatures overall. Climate *Change* is a more accurate description of what we are witnessing.

✈

This is certainly true in the areas I was operating during summer 2018 - Eastern and Southern Mediterranean. We saw plenty of change to the normal hot and dry forecasts which were expected between May and October. Why is the weather so important to us? Primarily because it affects the way in which we make decisions daily. For every flight that launches, there will be a decision made as to how much fuel to carry. For commercial IFR operations there is a minimum amount of fuel that must be carried legally. This will include fuel for possible holding, transit to alternate airfield and an amount to cover unforeseen contingencies.

It's a simple and well-tried formula that's been around for years. On top of this fuel however comes the discretion of the crew, or more correctly that of the aircraft commander to carry more fuel. It is often said that "fuel equates to thinking time" and this is certainly true when a flight has some form of interruption which prevents its timely arrival at destination. Having an 'elegant sufficiency' as I once heard it referred to, means that the decision-making process will not be adversely affected by time pressure – in theory. We must bear in mind that every extra 100kgs of fuel onboard the aircraft, costs extra fuel (burned) to carry it. If the airline is running 800,000 sectors a year, then this all adds up, which is why companies are keen to operate with a minimum fuel policy where it is appropriate.

✈

For me as a Line Captain, I believe I have detected a change in the attitude to the weather in the flightdeck community these past few years. It might be subjective, but these days it seems whenever I report for duty, the FO is ready and waiting to give me the direst prognosis. They usually start with something like "We WILL NEED EXTRA fuel for the *thunderstorms* today!"

Bear in mind that 90 percent of the time, the captain will already have checked the weather in advance of the report time. So, warning the other pilot of bad weather in the forecast should be a redundant action. Thereafter when we analyse the weather together, it is often the case that those 'big, bad, *thunderstorms'* are simply forecast Cumulonimbus (CB) activity. Remember TSRA (Thunderstorm with Rain) is what we are looking for in a TAF if we are going to refer to thunderstorms – if it's not there... well; we can't invent it. To see PROB40 CB (Cumulonimbus) is not the same as PROB40 TSRA.

Also, we should bear in mind the probability factor in the forecasts. If something is forecast as PROB30 (i.e. a 30 percent likelihood of occurring) then we can disregard it at the planning stage. While doing so, we should consider the previous forecasts and 'actuals' - the METARs. It could be that we can see a pattern of weather developing at the destination even before the next TAF is published. I recall one instance of this recently while dispatching before dawn from Burgas in Bulgaria. The forecast was for temporary periods of fog (TEMPO FG) at our destination airport in Belgium during the night, which had not happened. This was evident by looking at the METARs since midnight. Even now, the visibility was still 4km pre-dawn with scattered cloud – it was possible for us to go along with the latest forecast that stated that the fog was clearing up. It did not appear on the new one at all, just some mist. "Well we can handle a little mist..." I recall saying dismissively as we made our fuel decision. We could carry minimum flight-plan fuel with no

Active Thunderstorm off the coast of Sicily Summer 2018

problem... Although there was a little nagging suspicion at the back of my mind. That temperature and dewpoint on the latest actual were too close for comfort really – just 2 degrees apart - and there was a light breeze.

Remember your MET Theory Studies for the ATPL exams? Light W/V, typically 3-8 knots, with clear skies and a close temperature/dewpoint – ideal conditions for the formation of radiation fog. Well it hadn't happened on this occasion, so after briefing the rest of the crew, we headed for the jet. Onboard and we were setting up for the departure, the fuel truck had almost finished and my colleague was busy preparing the flightdeck to fly 'putting all the switches in a pleasing position'. I was still nagged by the thought of that temperature/dewpoint split and so I turned my iPad back on again and did a quick check on the very latest actual. Yikes! FOG!! "Damn! It's 200 METRES IN FOG!" I stated loudly and then *"We'll need more gas, don't let that fuel truck go!"* Meanwhile my colleague wound open his DV window and hollered at the refueller who was in the process of disconnecting. Of course, I was most apologetic when I got down to the fueling position under the wing and the bowser driver gave me a look like, 'Do you SERIOUSLY know what you're doing?'

I was disappointed in the first officer; in that he hadn't spotted it. At the briefing he had simply gone along with my decision making, although I could tell he was expecting us to carry more fuel 'just in case'. The 'just in case' philosophy, is about as useful as carrying extra fuel 'for the wife and kids' – it's simply not scientific. We should assess what is possible and be aware that the forecast weather conditions can be BETTER or WORSE than predicted. In this case, although the TAF covering the night period had been pessimistic and the FG did not come; suddenly nature had reversed her decision and brought the fog to the airport. This would mean Low Visibility Operations (LVO) approach with an automatic landing – that always requires more fuel. Not only that, but we should expect possible holding while airborne, due to the limitation on the number of inbounds which can make an approach in each time period.

Clearly, experience plays a part in the decision-making process. I was amused to hear one of my younger colleagues announce recently that he felt it necessary to "take extra fuel for the crosswind" I shook my head slowly and said to him, "Please show me where it says THAT in the manual?!" The forecast wind conditions for the time of our arrival were nowhere near the limits. Perhaps if he had flown in the Far East during the monsoon season his attitude would be different, or maybe in Africa around the Inter Tropical Convergence Zone (ITCZ). This is Europe, there are hundreds of airfields and options available to us all the time. It is not the Bay of Bengal or the GAFA*, so we should not overreact when we see rain and clouds. It is common to see red centres to build-ups on the weather radar here in Europe, but nobody is going to deliberately fly through one - or try to fly higher over the top of one.

My philosophy these days is that we should neither 'hate' the weather nor be afraid of it. It is what makes our job interesting! It provides mental challenges and sometimes physical ones too. Landing in turbulent, gusty crosswind conditions is something that the autopilot simply cannot do – for that alone we should be grateful. In the words of the song, we should Bless the Weather.

blame. You see, historically we have played down the dramas and difficulties of our occupation in line with the more important message that flying with the airlines is so safe as to border on boring.

easyJet A321 Captain & Mentor – Zoe Ebrey

We have many jokes on the subject. Like when we describe what we do for a living as: "Hours of utter boredom, punctuated by moments of pure terror!" For sure in the past couple of decades, the airline companies themselves have eagerly conspired to play down the role of flightcrew, not only to fit better with the modern corporate egalitarian philosophy, but ultimately to reduce the terms and conditions. Hence, we are now considered 'glorified intercontinental bus drivers'.

✈

There are parallels with bus driving of course. Both the airliner and the bus are autonomous vehicles in which the driver/s sit ahead of a large group of passengers travelling behind. The journeys undertaken are done so in accordance with the company's planned schedule and route structure with an emphasis on timely performance and safety. They are both 'public transport' operations that operate in a structured, regulated environment.

Absolutely ***everything*** to do with public transport exists in a regulatory framework. A large part of the challenge of our job then is compliance with the regulations. Often it is not enough to simply meet or match the standard required, but we aim to exceed those standards. It is true to say that ours really is a profession in pursuit of excellence. Let's face it, we have the highest motivation of all to be as good as we can at what we do – our own skin! We really do have 'skin in the game'. Not just our livelihoods, but our lives and those of our customers are at risk every time we go to work. We never forget - 'The pilot is always first on the scene of an accident'.

The daily obstacles that face us are many and varied. They come at us so thick and fast sometimes, that we must prioritise our responses to deal with the most serious first. For example, in the aircraft simulator we will be presented with a range of problems to solve as a crew and regardless of our individual strengths, we learn at an early stage that we must work well as crewmembers, so the team will succeed.

The Multi-Crew concept has been developed over the last 40 years or so as a distilled product of the knowledge gained from so many previous tragedies. Aviation history books are full of reports of hull-loss accidents which were the result of crew communication breakdown. I am pleased our industry is a world leader in human factors research and evidence of this is the concept of Crew Resource Management training for all personnel involved in the operation of airliners. Other occupations such as surgical teams in the medical profession are benefitting from the pioneering work done in commercial aviation.

Periodic skills testing is a big feature of our world these days. Every six months airline pilots are required to demonstrate their proficiency to an approved examiner in an Operator Proficiency Check (OPC). The check itself lasts for a minimum of four hours in the simulator, (maybe six if you include briefing and debriefing), but the prior preparation for the exam is intense. Candidates are encouraged to study for weeks in advance of the event. It is hard to describe the focus and energy expended in one of these checks. It is enough to say that after the successful conclusion there is huge relief on the part of the pilots. Not only that, but also there is a renewed confidence that they have passed the test and been cleared to operate on 'The Line' again.

On 'The Line' is where it all happens. It's all real here. No two days are ever the same and the perfect flight does not exist. Quite often in a large airline, the people whom we meet in the crewroom prior to departure are total strangers to us. There is a short time that could be called "The Golden 5 Minutes"; when we have the chance to brief as a crew together and share our knowledge of the challenges facing us in our assigned duties. After the introductions, we discuss the weather, airports, passengers, known technical problems and their effect on us.

Additionally, there is an SEP (Safety & Equipment Procedures) question and answer session between the crew to ensure their essential knowledge is up to date. It is at these times when you really appreciate the importance of Standard Operating Procedures. Complete strangers coming together to act as a coordinated team performing highly technical and complex tasks. Not just on one day, but every day they fly. It always amused me when I observed an airliner simulator being used to 'test' people on TV. The challenge was to carry out a safe landing on the runway without going off the side, or off the end. Not all were successful of course, some of the contestants lacked the necessary hand/eye coordination skills to complete the task, even with good visibility and no turbulence. The main thing that got to me though, was that they were only required to land safely once! Airline pilots must do it every time and often several times per day. If we were to fail, there is no electronic message that comes up on the computer screen saying; *"GAME OVER... Insert New Licence to Continue."*

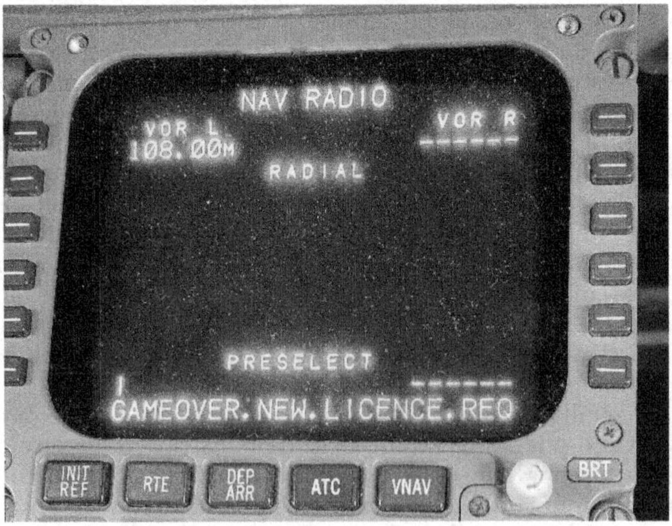

Naturally there is plenty of repetition in what we do. This is important, because having routines to follow ensures that the product is consistent. In this case our 'product' is the safe execution of the gentle art of aviation. The defeat of gravity is a hard-won prize and something that humans have dreamed of for centuries. We are blessed that our generation - and only one other before us - has been able to move across the surface of the planet routinely at 500 mph. All the ancestors before that had to crawl... There is a famous Isochronic map from 1914 by John G Bartholomew showing in 'numbers of days', how long it took to travel to various parts of the globe from London. The USA was 5-10 days away (ocean going liners), India was 10-20 days travel and Australia was 30 days plus, the contrast with modern airline network journey timing is stark.

✈

The development of the travel industry in the jet age can be compared to the advance of the internet and its subsequent effects on our lives. Nobody refers to these changes as tedious. The IT industry is racing headlong into the future, largely to the benefit of mankind. Similarly, aviation continues to evolve and with each breakthrough flying gets safer and more efficient. There are exciting prospects for anyone joining the airlines now. Adverse descriptions of a "Career of Tedium" may be confidently ignored.

✈

BACKSTORY: Back in 2019 I wrote this honest reaction to the unwise words of a journalist who clearly knows nothing about our industry. I feared this ill-informed view would dissuade more young people from following their hearts and entering a great career.

As I try to portray in my stories about what pilots face every day at work, our job is far from being dull. Career of Tedium? I flipping don't think so!

✈

me to make the R/T call, “TELL THEM we’ll be Ready on Reaching!” and I did not demur. I thought about it briefly and in a more enlightened operating culture I would have suggested to wait until we had the definite ‘Cabin Secure’ from the Senior; but this autocratic captain liked to deviate from Standard. If I had challenged him at this critical stage of flight, there would have been an unnecessary confrontation, maybe that would be bad CRM?

Ready for Departure? Keep it Standard...

Not a good thing just before we go flying. 'After all', the thought went through my head, 'the worst that can happen is we block the runway waiting for Cabin Secure and perhaps another aircraft has to Go-around...' Well guess what? That's *exactly* what happened! As we lined up on the threshold, we heard another airliner being given the instruction, "XXX, fly Minimum Approach Speed, one to depart ahead!" and then immediately, *"Break; Brunei 123. Clear Takeoff - landing traffic at four miles..."*

To which my captain said, "...tell them, we need another minute for the cabin...!" I did so, but what we were unaware of at the time, was the senior was standing in the aisle of the aft cabin having difficulty with a problem passenger. No way would we be getting Cabin Secure in the next 60 seconds. The controller was quick off the mark though, give him his due.

"Brunei 123 Takeoff clearance CANCELLED! HOLD POSITION! BREAK BREAK XXX! (traffic on finals) GO-AROUND! I SAY AGAIN GO-AROUND! RUNWAY BLOCKED!" Let me tell you, it's not exactly a calming sensation to know that a Boeing triple seven is carrying out a Go-around just a couple of miles behind you and with only a few hundred feet to spare. We heard the roar of their engines as they climbed over the top of us and we watched the landing gear retract as they disappeared back into the low cloudbase. Phew!

Very soon afterwards we got the call from the cabin and we were good to go. Now all we had to do was file the Air Safety Report after landing. It was a good lesson for me, even though it was years ahead of first officers being encouraged to 'advocate' their opinions of the operation to the captain. This event brings us neatly to another KISS that I think of more and more these days. ***Keep It Standard Stupid...!***

Let's face it, if we had adhered to the SOPs for our airline, then we would not have accepted the 'Line-Up' clearance. And we would not have ASSUMED that the cabin would be secure for departure in time for takeoff clearance. There are many examples I can think of when pilots deviate from the SOP in the interests of what they consider expediency, but then it all goes wrong. Don't take my word for it – there are several studies of commercial aircraft accidents where deviation from the SOP was a significant causal factor in a hull-loss.

✈

These days I fly with the next generation and at the start of each preflight briefing we are encouraged to 'verbalise the threats' as part of modern TEM (Threat and Error Management). This helps us to anticipate/mitigate possible threats and errors that could adversely affect the integrity or safety of our operation. Sometimes the FO will say, "Well, quite frankly... I have flown out of this airport so many times, I don't know what to say..." to which I reply, "How about 'complacency'...?" For sure that IS a major threat in an operation that can be repetitive, somehow, we must treat every flight as special and new despite the familiarity of the locale.

Often, I see pilots over-complicate the operation. For example, they may go to great lengths to build complicated arrival flightpaths in the FMC (Flight Management Computer). This creates a load of confusing magenta spaghetti on the moving map display. Bear in mind the first KISS. The simpler the better – plan on the most likely arrival routing, considering the 'shortest possible ground-track to touchdown'. Always be open to the idea that ATC may shortcut you during the initial or intermediate approach phase. Avoid being suckered into a High Energy Approach (HEA), to the runway with possible disastrous consequences.

Over-complication of the flying job actually creates stress which we really could do without. I recently flew for an operator which majors on ridiculously long-winded briefings. Seriously, they have an incredible grip on the trivia. One line training captain whom I flew with recently spent a full 20 minutes talking while giving me his descent and approach briefing for an airport with which we were both familiar. Meanwhile top of descent point was imminent and we needed to focus on the actual flying of the jet in my opinion. It is a cultural thing at that airline where they seem to spend too much time on a policy of CYA – (Cover Your A***).

So - my advice for a prolonged stress-free life? Keep it Simple, Keep it Standard and wear a clip-on tie!

BACKSTORY: I really do wear clip-on ties. I started wearing clip-on ties in the early 80s while I was working in psychiatric hospitals as a nurse, but that was for another reason! Seriously though, I'm a great believer in keeping things as simple as possible while flying aeroplanes. Don't get distracted by trivia – like the crew who flew into the ground while focused on changing a tiny little light bulb in the cockpit instrument panel...

11 THE MINISTRY OF GREY AREAS

I recall in the early days of my commercial flying career we used to refer whimsically to a mythical entity called - "The Ministry of Grey Areas." This was sometimes suffixed with another word in brackets, to identify which department of 'the ministry' we might be referring to. For example, The Ministry of Grey Areas (Technical) was clearly devoted to engineering and technical matters. You might say it was an immature way for trainee professionals to talk about the regulating authority – in those days the UK Civil Aviation Authority - however it is also evidence of our frustration with an organisation that was not without its faults.

The primary mission of a Regulator is to regulate – that is, to apply the *regulations. This may seem a completely obvious statement, however as recent events have shown, there appear to be grey areas in the light aircraft charter regulatory environment that require clarification and/or enforcement. In the accident of January 2019 which claimed the lives of footballer Emiliano Sala and pilot David Ibbotson when their aircraft crashed into the sea near the Channel Islands there are suspicions as to the legality of the flight itself.

✈

There is doubt about whether the flight should have qualified as a private flight – i.e. undertaken NOT for hire or reward. There were commercial considerations involved for sure. The passenger was not the owner of the aircraft, he was a celebrity sportsman being transported at the behest of a third party or agent. The pilot was not the owner either and in fact the true ownership of the airframe is further complicated by the state of registration – in this case USA. It is legal for 'N' registered aircraft to operate privately in states outside the US, however for a commercial flight to be undertaken the FAA would require that the aircraft be flown by a licensed operator. The compliance with all the regulations necessary to set-up as an Air Taxi operator under FAA regs would be prohibitive for companies in Europe.

The pilot himself may or may not have been appropriately licensed/qualified to operate the flight, given that it appears to have been made in IMC (Instrument Meteorological Conditions) and/or at night. No doubt the focus of the investigation would be on who knew what and when. In these days of public liability and litigation, the audit trail will not be hard to find.

✈

For sure the regulations are there for a reason – primarily to keep

12 ADORABLE ANALOGUE

There was a time when all the flightdeck instruments that we looked at were analogue. Pointers on dials, sweeping past numbers to indicate height, speed and heading. Not only these, but also, fuel capacities, temperatures, pressures, they were all clock type displays. There was a certain elegance and completeness in their arrangement across the cockpit instrument panels. Instruments? Yes, they were often the final link in the information chain that began at the sensor, somewhere else on the airframe. Whether that was a temperature probe, or a gyroscope referenced to magnetic north, they all came together in the display of information to the pilots.

Of course, not all aircraft cockpits were designed as complete and finished products from the beginning. I recall once looking into the front office of a 1960s Buccaneer at RAF Lossiemouth. The year was 1983 and I was accompanied by my RAF QFI (Instructor) at the time, who was an ex Banana-Jet driver. I sat in that tight little space and he showed me how the cockpit had received odd updates in instrumentation over the years. This included an additional airspeed display that ran across the top of the cockpit coaming (glareshield). To be honest it looked a bit gimmicky and for all the world like the sliding scale speedometer in the Rover 3500 of the time. Those early attempts at being futuristic didn't always impress us.

Even when aircraft designers had the technology to present information to aviators in digital format, it appeared they were reluctant to do so too quickly. I am thinking particularly of those first glass cockpit Boeings. In the early '80s when the 757 rolled off the production line in Seattle, there was huge excitement in the aviation press about the new "Glass Cockpit" airliners. After all, the primary attitude and navigation instruments were now TV screens with computer generated displays. We all talked about making the transition "from steam to glass." Funny old thing though, when we actually got in there, the TV screens were accompanied by some very traditional looking airspeed and altitude dials. What we learned later while doing the technical course, was that those analogue instruments that appeared to be 'normal' were *displaying electronic info from the Air Data Computers!* Now THAT came as a shock. The age of the silicon chip had almost arrived.

Only the Standby Instruments were separate. I am uncertain as to the reason for the ASI (Airspeed Indicator) and altimeter not being included in the original EADI* displays. Maybe it was a step too far technically, or they didn't want the whole ensemble to appear too 'radical', such that pilots might have trouble adjusting their instrument scan - it could have been both. Whatever, none of us was unhappy about the comforting feel of those round dials. In fact, when the first speed-tape ASIs came in some years later, there was plenty of concern from the Human Factors experts about the possible misinterpretation of the indicated speed.

They should not have worried, by that time we were happy about the concept of digital. In the early days of my training as a First Officer on the 757, I vividly recall a Senior Training Captain advising me to make the *"Eighty Knots"* and *"Vee One"* calls as Pilot Monitoring, by reference to the digital display on the ASI. This was instead of trying to judge the precise moment when the needle swept past the printed number on the outer scale of a round dial. As he explained to me, in the event that there was an incident or major event during the take-off roll; at the subsequent investigation, my voice would be clearly recorded on the CVR at the exact moment of those critical speeds... this was sound advice, if you pardon the expression.

The Boeing 757 was the first time I had come across a totally digital clock in an aircraft too. Now here was a thing. The trouble with the early 757 clocks though was that they didn't keep the most accurate time. Worse still, there were two of them, one on the left (Captain's) side and one on the right. It was quite common to find that one or both were wrong by a minute or two. Not only that, but they were often mismatched between each other by more than a few seconds. Of course, that's the problem with a digital clock, either it is correct to the minute... or it is not. The later aircraft were much better and the 737 models of the 1990s were spot on.

It is interesting to note that even when the manufacturer gives the crew engine gauges that are depicted on computer screens, they still include a needle and dial display combined with the digital. It seems the engineers are loathe to dispense with analogue completely. And that's where analogue has a place even now. It is quicker to compare two small dials with needles, than to read two sets of digital values and interpret them. Further, it is more likely that you be able to 'call' the difference correctly first time. That being said, there are very few occasions on an airliner flightdeck when the pilots are required to make a snap decision. The old maxim of .”..sit on your hands and think about it” (before doing anything), still applies.

As you may have already divined from the title of this article, I still have a strong affinity with analogue instruments, even though I appreciate the technical beauty of the most modern displays. Back in 1999 I recall our first encounter with the B737 NG (Next Generation) cockpit. It was a revelation to us that now it seemed our manual flying skills had deteriorated overnight. The reason? Because the altimeter readout was digital to the nearest 10 FEET! All our minor 'height-keeping' errors were magnified... Whereas before we might have talked about flying to the nearest 100 feet, now we were aiming for the exact nominated altitude, plus or minus zero! But I'm not complaining, our accuracy did improve as our manual instrument flying skills developed with practice on the new equipment.

Summing up the beneficial effect on our flying accuracy, I guess you could say some things don't change. Flying aeroplanes for a living is just like it's always been. If you give us the tools, we will do the job.

✈

*EADI – Electronic Attitude Direction Indicator. The term used for the primary flight instruments combined into a flat screen display in the first generation 'glass cockpit' airliners.

BACKSTORY: When the first Boeing Next Generation instrument display arrived in easyJet in late 1999, it took our breath away. However there is a lot to be said for the good old dials – especially if you are in a rush for information. The brain can assimilate the position of a pointer on a gauge quicker than assessing a digital readout.

✈

13 BLACK AND WHITE

We work in one of the most highly regulated industries of all time. Of course, there are overriding reasons why this must be so. Younger colleagues often ask me why they must memorise or study so much written material. There are a lot of ways to answer this question. Nowadays I cut straight to the chase in my reply,

"It reduces the grey areas. They want it all to be black and white." This is the crux of the matter. As operating crew, we must have clear guidance so we can manage every situation safely and effectively. An easy example is when we encounter an obvious technical defect.

It was the first flight of the day, we arrived at the aircraft, a Boeing 737-800 just before dawn to carry out a flying duty of four sectors. The ship was cold – in other words it had been completely shut down after arriving the night before. The engineers had carried out their checks and signed it off as being serviceable. As the first officer would fly the first sector, he went to open it up while I carried out the exterior check.

Note, that not all airlines are the same. Some stipulate that the aircraft commander carries out the external inspection, while others simply state it is the task of Pilot Monitoring. While working for Alitalia, back in '96 it was always *"Il Comandante"* who was nominated by the company to conduct the external checks. Occasionally I would fly with an FO who volunteered to do the walkround, whereupon I would politely remind him that with Alitalia it was the captain's duty.

On this morning, I started at the front left-hand side of the 737 and commenced to move clockwise round the ship. Before I dipped under the nose to have a look in the nosegear bay, I reminded the FO to switch on the hydraulic pumps as soon as possible when he got to the cockpit, this would ensure I could easily spot any leaks. After inspecting the Number 2 engine, I moved outward towards the wingtip and noted that the starboard navigation lights were out. There are two bulbs in each navlight housing to prevent a single point failure. I assumed that the FO had not yet switched the navlights on and continued my walkround.

When I saw that the tail and port wingtip lights were working, I returned to check the starboard ones again. They were still dark – nothing - both bulbs were out! It was still a couple of hours until daylight - we were due to depart before dawn. *Yikes!* I moved quickly to the steps, yelling at the Dispatcher to hold off on boarding the passengers. Entering the cockpit, I checked that the navlights were switched on and told the FO that we had no wingtip position lights on the starboard wing.

very slight clicking/interference had not just occurred overnight. Probably the aircraft had been flying like this for some time. The machines in this 737 fleet normally achieved eight or ten sectors per day, every day throughout the month. So, the decision was *"We'll take it"* and after completing the before taxi checklist we called for taxi clearance.

In fact, the aeroplane operated perfectly normally, however I was still puzzled as to why none of the previous crews had snagged it. When we arrived at our destination after the three-hour flight, we checked the controls again and it was still there – we had not imagined it. The funny thing was that there was a difference in what was heard/felt between the control wheels. If the full and free control checks were done with forces being applied from the right-hand side of the cockpit, it was perfectly normal. But it was not the same from the other side. Then it struck me.

Of course! Usually the FO would be Pilot Flying for the first sector of the day and therefore THEY would carry out the full and free checks before the first flight. The Captain would then do flight control checks for the second sector as PF and quite possibly they might notice a very slight clicking/restriction of movement at just over 75 percent, but by then the plane had flown and behaved entirely normally. This would put a different complexion on things. For sure this problem was not black and white.

On our return to home base we asked for the engineers to meet us on arrival. It was time to get this one solved. Again, the machine operated normally during flight, however that slight control restriction was still there as if a cable was binding slightly on a pulley at full stretch. I called Maintrol and wrote the fault in the Techlog with a full description of the symptoms. When the engineer arrived, I explained the problem to him and showed him the Techlog entry. For sure the problem was black and white now, no more grey areas on this one, although I could see that the engineer was not convinced.

He said, "but you would never use 75 percent of aileron/control wheel input inflight?" To which I replied,

"THAT'S not the point. The question you must ask is this. Are the flight controls 'full and free' when you move the control wheel in all axes to the full deflection? If not - then it must be addressed. All the other 737s ARE full and free!"

I knew that the aircraft was now grounded until they carried out rigging checks on the flight controls and that would not be a short job, however it was at a maintenance base for the airline with spare ships available. Also, it was not at the height of the summer with a back to back programme and nil slack. It was a good time to get it done. I never did find out what happened to it but was reassured that the maintenance team would fix it. The standard of maintenance at this airline was extremely high with hardly any aircraft flying with officially acceptable deferred defects.

I guess the moral of the story is this. Grey areas do exist in the operational flying environment and sometimes you end up scratching your head, wondering what to do? However, when information is passed on in writing things become black and white. Certainly, this is true in the relationship between flight crew and engineering. We communicate via the Aircraft Technical Log which is considered a legal document. As pilots flying the line, we must try not to diagnose the problem, but we simply describe the symptoms.

At the same time, we should resist the temptation to write up trivial faults with aircraft. I once knew a pilot whom the Base Engineers referred to as Captain Snagalot and you really wouldn't want to have that nickname.

BACKSTORY: Both of these B737 tech issues occurred at Alicante Airport within a couple of days of each other. Which was quite odd, because usually the aircraft were faultlessly presented for flight with nothing in the Hold Item list (ADD) part of the Techlog.

I know the maintenance guys were annoyed when I called directly to the Maintrol hotline to report the navlights issue – but what else could I have done? They had just completed the daily inspection which is good for 48 hours and that list includes all the lights.

There was no way I wanted a delay being attributed to us. The engineers had signed it off and gone home.

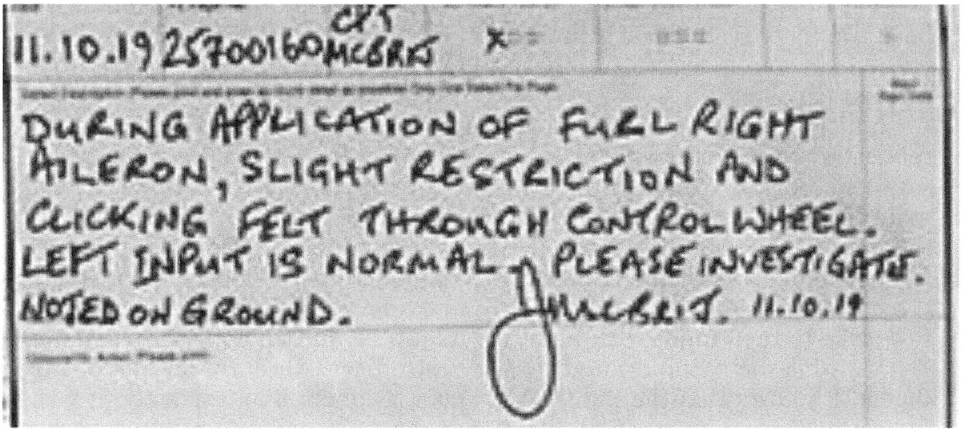

11.10.19 25700160 CPT MCBRIJ

DURING APPLICATION OF FULL RIGHT AILERON, SLIGHT RESTRICTION AND CLICKING FELT THROUGH CONTROL WHEEL. LEFT INPUT IS NORMAL. PLEASE INVESTIGATE. NOTED ON GROUND. J MCBRIJ. 11.10.19

Captain's Techlog entry for control restriction – Engineer said "you never use full aileron"

"THAT'S not the point... all the other 737s ARE full and free!"

14 BELIEF

Long before I started flying myself, I knew that pilots had to believe their instruments. Whatever they felt or however they reacted to any given situation, the primary source of reference in their world was their flight instruments. The reason I knew this was because my friend's father had been a pilot all his working life. He was adamant – "Always, always, always, BELIEVE your instruments!" He had been trained by the Royal Air Force in WW2 and had then continued his aviation career afterwards with a series of commercial operators for another 30 years. Clearly his initial instrument flying training had been effective, he never forgot the primary rule of I.F.

Of course, now I realise he was talking about flying in cloud without external visual references. Not only that, but he also referred to prioritising the information being received by the eyes over that of other possible contradictory feelings sensed by the body. For example, the Somatogravic Illusion of pitching up as the aircraft accelerates during a Go-around which has resulted in some pilots pushing forward on the control column and flying into the ground.

That being said, "believing what the instruments are telling us" comes with a note of caution. At all times we should be prepared to cross-check and compare the indications to preclude the possibility of instrument failure. One of the principles of modern aviation is that there must always be at least one back-up system and quite often two. In the case of the hydraulics, we have two main systems on airliners and even a standby system in case both fail. It is of course possible to fly and land the jet with just one of the three systems.

Similarly, in the case of the flight instruments, we have a complete replication on the other side of the cockpit. Not only that, but there is another set of standby instruments for use should both captain's and first officer's main instruments fail. The Standby Attitude Indicator is positioned just to the right of the captain's main AI where both pilots can see it. In times of total main flight instrument failure, the manufacturer obviously intended for the captain to be able to manually fly the aircraft. He/she would fly with reference to the Standby AI, along with altitude and airspeed instruments nearby while being monitored by the first officer. Recently on one of my flights, I noticed that the Standby AI was not functioning correctly. It was indicating three degrees of bank in the cruise and would not reset properly. I said to my colleague,

"That's a no-go item in the MEL*, we will snag it when we get in. It's the last flight of the day fortunately so the engineers can replace it overnight." Now the interesting thing about this statement was I already KNEW it was a serious defect and therefore not possible to dispatch with it for flight in IMC (Instrument Meteorological Conditions) or at night. You see it is not a requirement for operating pilots to 'know' the contents of the Minimum Equipment List by heart, but there are some defects which stick in your mind. My young colleague said,

"How do you know it's a stopper?" to which I replied, "I had the same thing on another flight, same aircraft type, about 15 years ago." Then I explained the circumstances.

Bill Wilkin – "Always, Always, believe your instruments"

"We HAVE to believe that the engineers know what they're doing" – they had diagnosed it from afar as best they could. Then came the words I longed to hear...

Captain it's FIXED! You are good to go, I will just do the paperwork and you can close the doors!."

I looked at my watch, we MUST be off the blocks in 5 minutes or else! We pushed back inside three minutes and landed in Luton less than two hours later – job done.

*MEL = Minimum Equipment List for commercial transport aircraft to dispatch. All possible aircraft defects are covered here, and their operational impact is considered.

BACKSTORY: it was ex RAF Flying Officer Bill Wilkin who impressed on me to "Always believe your instruments!" He was the father of a very good friend of mine. His RAF and civilian flying experiences would have been enough to fill a book on their own. He was an intelligent man and a gifted engineer.

Since having the issue with the Standby Attitude Indicator and collapsing gyro inside it, I carefully monitor the condition of that instrument. I inform Maintrol as soon as I see one that looks like it might be on its way out. It's a real stopper if it happens down-route!

The funniest thing about our whole Barcelona episode was that one of the passengers was caught smoking in the lavatory on the inbound rescue flight from Luton. The captain arranged for the police to arrest the passenger on arrival in Barcelona before the crew flew on to Palma Majorca. We all thought this was hilarious of course – it made our day!

cabin and arrived at the rear galley – she was close to tears and her face was white as a sheet. She identified the offender. I held out my hand in front of him and requested his passport and boarding card.

He feigned not to understand and mumbled excuses in French, but I was insistent. I repeated my request clearly and loudly this time in French,

"Votre Passport s'il vous plait Monsieur, avec La Carte d'embarquement, MAINTENANT!" When he handed them over, I said very nicely, "Merci Monsieur. Please wait here this won't take long" and walked back to the forward door. As I descended the steps, I was pleased to see the handling agent had reacted quickly and the Gendarmes were there also.

I presented them with the passport and boarding card, informing them that this passenger, M. ________ had been aggressive and threatening towards cabin crew and refused to comply with safety instructions.

"He is denied travel and we request your kind assistance in removing him from the aircraft NOW!"

I half expected the passenger to resist when I went back down the aisle, however when he saw that I had brought Monsieur Gendarme with me he knew the game was up. He immediately began to get his belongings together to disembark the flight with his three children.

I felt badly that the children had to undergo the trauma of being offloaded by the police. One can only hope that it instilled in them the realisation that there are consequences for their parent's bad behaviour. His was not a good role model to follow.

The incident signalled the right message to junior crew members who must deal with the public every day that there is a limit to what is acceptable. They need not suffer rudeness and abuse by passengers during their working day.

From our perspective, sat in that little room at the front, the whole event was resolved inside 30 minutes and we got a new slot. I made the LMC** adjustment to the loadsheet without any regrets – 'Offload 4 Pax + Bags, reduced weight 245kgs" – In fact *it's a bonus* I thought, we will save a bit more fuel!

✈

*Pax = Industry shorthand for "Passengers."

**LMC = Last Minute Change, made to the Loadsheet for weight & balance of the a/c.

BACKSTORY: We bid adieu to this disruptive passenger in Toulouse as we departed for Fez. This passenger had also failed to read his booking conditions, because when he arrived at the check-in without having printed off his boarding cards, the airline then charged him another EUR 240. Ouch!

✈

were mercifully spared any accidents, there was a constant stream of incidents, some serious.

Well it was during this time of great strain, when we were without our Chief Pilot and the other two Regional Fleet Managers, my boss, Jim asked to see me. We had a meeting in one of the little rooms at easyLand. He looked exhausted and said so. He said he needed to take a long weekend with the family to decompress and focus on them. I assured him that I was happy to fill in and promised not to bother him while he was having time out. He briefed me on all the current issues, he forwarded his mobile calls to me and passed me the keys to the office. Finally, he made sure I had a copy of the company emergency response manual. So that was it, Friday afternoon and he was gone.

"See you next Tuesday" he said as he was leaving.

My first visit was to NMC to make sure the fleet was running well. Fortunately, it was. Virtually nil technical issues: no more than average crew problems and the weather was forecast fine for the weekend throughout Europe. Also, the team looked in control, there was a quiet hum from the air-conditioning for the computers and conversation was peaceful and unstressed. I was happy.

✈

But now it was Saturday evening.

"James. It's Sir Richard Branson"

I thought to myself, *'Oh F***!'* and put my head in my hands. Richard Branson? The founder of Virgin Atlantic and the most high-profile person in the British aviation industry. *Dear God, why does it have to be him?* Of all the things I could have predicted, this was not one of them.

Somebody had to decide within the next 10 minutes as the flight was due to depart. I asked for all the details and Johnny said.

"He, Sir Richard, is asking for us to let him travel in the flightdeck on the jumpseat. The captain is okay with it, but he won't do it without permission in writing from the Chief Pilot... "

Or in this case, the deputy Chief Pilot - me. I promised Johnny I would get straight back to him and put down the phone. What would my boss Jim do? Well... he would say 'No' straight away I reckoned. Simple as that. He had a permanent position to protect as a CAA Postholder. Yes, whichever way you looked at it, if he could not avoid the decision entirely, he would say 'No'.

Although, if I called him up and explained it all, he might say 'Yes', *but then his job would be at risk.* If we say 'No' right now, they may try to strongarm one of the fare-paying passengers off the jet just to board a VIP. Imagine the bad PR of that one? Then I considered the big picture. We were under a lot of pressure in flight operations and I knew the line crews were feeling uneasy about a lack of leadership in the office.

The Chief Pilot had been very popular and was sorely missed. Ditto my two colleagues who had left for the competitor airline. Losing Jim too, based on his making a bad judgement call would be a huge blow. *But me? Nobody would worry too much about the Deputy getting shot, would they…?* I took a deep breath as I pressed the button for NMC…

"Hi Johnny. YES, it's on. You have my authorisation, I will send it to you by email now. Forward it to the captain of the flight. Sir Richard owns an airline and he's travelled in the cockpit many times, in fact he's a pilot, possibly with a commercial licence. We cannot afford the bad publicity of dragging off another paying customer to facilitate a millionaire celebrity."

I shrugged. *No drama if they remove me from my position. I will have taken one for the team right?*

✈

On Tuesday Jim returned to the office bright and breezy. Three complete days away with four uninterrupted nights' sleep had worked wonders. The bags under his eyes had vanished; he was focused and positive.

"Anything happen while I was away?"

I told him about the Saturday night dilemma and the rationale for my decision to authorise Sir Richard Branson to occupy a jumpseat on the flightdeck back to Gatwick. He was shocked, then realising the bullet he had dodged, he asked,

"Erm, no comeback yet then?."

"No. The flight left on schedule, got into Gatwick early and everyone was happy."

He was smiling as we walked out of the meeting room and I asked him.

"What would you have done if it had been your call Jim?"

"Oh, I would have said 'No' James."

Then it was my turn to smile and nod,

"Yes… that's what I thought you would have done."

✈

BACKSTORY: This happened during a tough time for easyJet. We were short of everything back then, crews, aircraft, ground personnel and managers. Everybody was working flat-out to make 'the dream work'. The growth of our airline from a fleet of four to 100+ in less than 5 years had not been in the least bit 'easy'.

17 ENERGY MANAGEMENT

When people ask what is the most important aspect of our job after Safety; I often reply, "energy management." They sometimes nod their head wisely and reply,

"Hmmm, I thought you'd say that." The fact that they don't wish to explore the subject further after this, indicates to me that maybe they don't really get it. They don't fully comprehend the correlation between 'energy' and 'efficiency'. The two are inevitably interlinked when discussing the operation of airliners. If we burn more fuel than necessary to get the job done safely when we fly, then we are negligently misusing our planet's resources.

✈

Of course, it is not just the fuel, but the time. If we manage to save time in the air, then our efficiency level increases - plus less fuel is burned going from A to B. Which explains why we are always asking ATC for direct routings and shortcuts. I recall 30 years ago thinking how very inefficient it was that we were required to remain on/in "The Airways" - tracking the centrelines of invisible paths between the radio beacons set up many years before. We zig-zagged our way through the skies simply because of protocol and out-of-date thinking. Technology had gone far beyond that.

Even back in the '80s it was possible to track between waypoints with our inertial navigation systems. Those waypoints had been defined simply by Latitude and Longitude. Even then those invisible corridors were archaic. Sadly, they were deemed necessary for political reasons. Finally, here we are, at the end of the second decade of the new millennium, thankful for GPS, being mandated to fly point to point (using Great Circle tracks) across Europe over vast distances. This new/old invention will, in addition to the advent of CPDLC. This is a computer data system inexorably replacing traditional ATC R/T methods, which will ultimately lead to more efficiency and fewer human controllers.

This probably explains the modern fashion in the London TMA (Terminal Manoeuvring Area), for controllers to assign more radar vectored headings, speed and altitude restrictions than hitherto. How often, when departing from a London airport do we say to ourselves in the flightdeck,

"WHY won't they just let us fly the SID...?" (Standard Instrument Departure) often it seems that our consequential and repetitive radar headings lead us to PARALLEL the SID routing! Forgive me for mentioning anecdotally, but it does seem that the London controllers are making work for themselves.

Commonly we find the frequencies working to capacity as flights join and leave with more vectors, restrictions and complex ATC instructions than ever before. You could ask,

18 THE ITALIAN JOB

As I walked into the art gallery meeting venue, I was conscious that I was a few minutes late; parking in West London is never easy. Neil (our boss) had already started speaking and his audience of 30 plus airline pilots hung on his every word. As usual he was eloquent, amusing and profound all at the same time.

"I am so pleased to see so many of you here for our reunion, thank you all for coming and thanks to Ivor and Phil for their organising efforts" he paused and beamed at us all. You could have heard a pin drop, total silence. So, with a twinkle in his eye he cleared his throat and continued,

"Well... I suppose I'd better tell you a joke." We all giggled; it was typical of him.

"It's about the day when Neil Armstrong landed on the moon in July 1969 and transmitted those words, '*One small step for man... One giant leap for mankind*', but he also said at the end '*Good luck Mister Gorski!*' Now the team at NASA in Houston Centre didn't know what to make of this, but as there was a slight lag, they quickly dubbed some static over this part and released the audio to the world. The thing is, that about 30 years earlier, Neil had been living with his family next door to the Gorski family in Ohio. In the back garden one evening, while he was only eight years old, he overheard a loud argument between the neighbours. It transpired that Mr. Gorski had been requesting his wife to perform oral sex for him. Allegedly she shouted, *"that's completely disgusting"* and vowed *"I will only do THAT for you, when the kid next door, walks on the moon!'"*

We all roared laughing. It was the first time I had ever heard the 'Good Luck Mr. Gorski' story. As with all wonderful occasions, this evening was over far too soon. At the end of the night I recall we were singing our hearts out in a very nice Italian restaurant. The song? ***'Volare'*** of course! You know it's funny, but I only recently studied the English translation of our beautiful anthem, as in those days we just sang the Italian version.

"Then suddenly I was rapt by the wind and I began to FLY in the infinite sky...." The lyrics are apt and perfect for expat pilots flying in Italy! But of course, you're probably wondering what this reunion was all about.

✈

I am sometimes asked, "Of all the many airline jobs you've ever had over the years, which one was the best?" Although I always take a deep breath and think, it comes back to the same answer every time,

"Without a doubt, The Italian Job!" It was a time when I flew as a captain on the B767 doing long-haul for Alitalia. Naturally it was a contract. There is no way Alitalia would ever employ a non-national if they could avoid it. So how did it happen? Back in the early '90s Italy's national carrier wanted to start a new long-haul fleet. At the same time they planned to finish with the McDonnell Douglas DC-9.

So they leased a fleet of Boeing 767-300ERs that would have the range and capacity to service their intended routes. Unfortunately, when the DC-9 crews heard about the plan, they decided to obstruct it. All of the DC-9 fleet personnel went on strike. Hundreds of pilots and cabin crew were sat around at home on basic salary, while the company struggled to get the new widebody Boeing fleet up and running.

Naturally the unions backed the crews all the way and Ali had a fight on their hands. The pilots told the airline it was not safe to fly across the Atlantic with just two pilots – they should have a Flight Engineer or inflight relief crew. The cabin crew refused to fly all that duty time without a large crew rest area. They demanded the last six rows of seats in economy were removed to accommodate it. The airline management board were horrified. They had already committed to leasing the airframes from a major Australian company and now their own crew were refusing to operate them. What to do?

This was before deregulation of the EU skies and before a harmonised (and interchangeable) flightcrew licensing regime. Another complication was that the leasing company decided it was too risky to have the machines registered in Italy (if they needed to be repossessed for example at a later date). They registered them in the UK as 'G' registered airliners. Now to operate them, they would need UK licensed pilots – can you see the problem?

Enter the UK pilots from the reunion. Lots of B767 type-rated crew with UK CAA licences headed to Rome and Milan in 1995 and '96. Fortunately for us 757 drivers, the 767 is the same commercial type rating. So we only required a short differences course and a couple of simulator sessions to fly them. In early 1996 I joined the merry band of brothers in Rome and I didn't look back. Due to the time pressure, there was rather a lot of expediting going on, if I recall correctly. But my greatest memory is of the fabulous esprit de corps. It was how I imagine joining the Foreign Legion might have felt. Not only that, but the terms and conditions were highly attractive. Seriously we felt like we were being spoiled.

Firstly, we were handsomely rewarded financially, then provided spacious, serviced apartments in Rome or Milan. They ferried us to and from the airport in limousines. This was standard Alitalia practice in those days, nobody parked their own vehicles at the airport. Cabin crew were all picked up from home by minibus, the pilots by a Limo. When we arrived at our destinations, the hotel reception desks would disburse cash in the local currency to spend as overseas living allowances. You couldn't possibly spend the generous sum on the trip.

G – registered B767-300ER on ramp at JFK

The feeling of just being there was incredible – it was like being in a Hollywood movie! Our bespoke uniforms were crafted by the best tailors and required at least two fitting appointments at the shop in the centre of Rome. Sadly, I arrived just a month too late to be issued the famous Alitalia greatcoat. Pilots would wear it over their shoulders in debonair fashion, hardly ever slipping their arms through the sleeves. Similarly, Alitalia pilots prided themselves on carrying, not wearing their uniform hats. I was told this was because the hat had so much gold on the front, they felt they looked like the *Carabinieri* and that was definitely NOT a good look!

Everyone wore sunglasses all the time it seemed and most of the crews were talking on their mobile phones while walking through the terminal, usually balancing a coffee and/or a cigarette with the other hand. The flying was the sweetest airliner operation I have experienced. Mainly I think because the crews were treated like royalty. Our flights were dispatched in the old-fashioned way with properly trained company Dispatchers working from their well-equipped operations office in each airport you went to. They were experts at reading upside down I noticed as they presented you with the flightplan, the weather and Notams.

Nothing was rushed of course, for some reason we ALWAYS had plenty of time. If the flight pushed back late... no big deal. Plus, all the aeroplanes were brand new! These beautiful ships were like new cars direct from the showroom, they looked and felt amazing. In addition, they still smelled like new, with no trace of "Eau de Boeing" that developed after a few years. The cabins were generously spaced out with 20 large business class seats and 214 economy class. The engines were General Electric CF6 high-bypass turbofans. I think there is a large fan club out there for these engines – they are so well respected and liked by all the flightcrews who operate them.

Oh, and I mustn't forget to mention the fuel. We were never limited regarding fuel uplift and we regularly shut down in New York with over 10 tonnes in the tanks having crossed the Atlantic. I know that might sound like a lot for most aircraft, it was still rather a lot even in a B767! "Il Commandante" always performed the walkround and the SOPs were geared towards the left-hand seat occupant being very much in charge. Of course, those of us who had flown in the modern UK charter airlines found this a little 'old fashioned' so we tried to break up the seniority gradient with task sharing and decision making.

The cabin crew were nearly all from Australia. Some of them had never flown before, but they all had something in common - they were fluent in Italian. A common dislike of their uniforms also united them. They were an Armani design that looked superb, but according to the wearers, were uncomfortable and poorly fitting. In fact, when we carried Giorgio Armani himself to New York in business class, the cabin crew told him exactly what they thought. To which he replied, "ALORA! My design was good ladies. But don't blame me because Alitalia went cheap on the material!"

Talking of New York, reminds me of the amount of time we spent there. In some weeks it was more time in the Big Apple than in Italy. The cabin crew would all make their beauty appointments for downtown NYC, some even enrolled in night-school classes there. The crew used to laugh and say,

"We're actually BASED in New York; we just do flights to Italy!" We stayed at the legendary St. Moritz situated on Central Park South – right in Manhattan. Il Commandante routinely was allocated a suite, with a lovely view over Central Park. Heaven! There were crew parties, crew dinners, shopping expeditions, sightseeing tours, concert/theatre visits, love affairs (of course) and everything else besides. Of course, the jetlag was not easy to cope with and one of our young first officers learned about life the hard way. He decided to take a walk in Central Park at 2 am one morning. He was mugged and lost his very expensive Rolex watch, but as we all said, it could have been worse.

So, there I was at the age of only 39 in an Alitalia captain's uniform on a wide-bodied jet. The stuff of dreams really. In the normal way of things, I would had to have been Italian at the very least to achieve this. Our happy team of contractors all knew each other well and there was a strong sense of 'one-mission' among us. Our routes included Accra, Boston, Chicago, Tehran, New York (both JFK and Newark). We operated all these destinations from our two bases at Rome Fumicino and Milan Malpensa.

In early '96 Malpensa had only just started to open up after some long delays in building the runways. Naturally, not everyone knew there was a small cadre of British licensed pilots flying for Alitalia and one of the threats was experienced very early on when ATC used to call us in Italian. Climbing out from Malpensa over the Alps, bound for JFK, we were often addressed as "Alitalia Sette Sei Zero Due..." We soon learned our Italian numbers so we could at least recognise our callsigns! An embarrassed radar controller then continued in English. So much for this being the standard aviation language for international flights.

Pilots experienced a few embarrassing moments as well. One of which was when one of the Rome based captains inexperienced with B767 operations, elected to return to the airport after one of the engines appeared to be low on engine oil. The General Electric CF6 engines did have a little idiosyncrasy in that one of them would 'hide' oil for some reason. It was normal and we all accepted that after take-off, we'd see a lower indicated oil quantity on one side than the other. There were no other abnormal engine indications and all the aircraft did it. In the early days however, this information was not widely known, hence the turnback.

My own cringeworthy event occurred while we were taxiing after landing in New York JFK. As you can imagine, there are literally miles of taxiways there and they always seemed to be rebuilding the airport. That's not an excuse however, because it was my fault when I took a wrong turn on the taxiway. Suddenly I had the cold shock of realisation – we were taxiing down a dead end as I had turned right one exit too early. My colleague in the other seat quickly confirmed the error and was gracious enough to apologise for not following our progress on the taxi chart as well as he could have. Now we had a Boeing 767 in gleaming Alitalia colours, with that big 'A' in red and green on the tail heading the wrong way! Aaargh! I could have screamed, but

I held it in and breathed deeply.

There was constant chatter on the ground radio frequency from all the other traffic. Any second, I expected the controller to call us and ask what exactly we thought we were doing. Just then I spied an intersection - JUST wide enough for us to turn. So, I slowed the big machine right down, got very close to the edge on my side, then wound on full steering lock with differential power on the outside engine. We did a 180 and in less than a minute we were heading back the way we had come. Now on the original taxiway, we made the correct turn, and nobody said a thing. What some of the passengers thought, I don't know. I was so relieved! I am pretty sure we were not the first nor the last to take that wrong turn, oftentimes it was controlled chaos there on the ground first thing in the morning.

Not just Kennedy Airport, but Newark also was extremely busy, and we would load extra fuel for taxiing pre-takeoff. There were huge queues of airliners in those days waiting to get airborne and it was not uncommon to have at least half an hour taxiing after you pushed back off the gate. The longest wait was 55 minutes taxiing prior to take-off. We even heard a Virgin Atlantic 747 "Request departure, FUEL PRIORITY!" which made us laugh. 'Fuel Priority' for arrival is something that all airports will respond to, but for departure? I think they were joking, but maybe not.

While on the subject of humour... there was plenty of that on our fleet. I cannot remember a single trip when the morale was not sky high. Once we were one cabin crew member short to depart from Milan for JFK. Our crewing people called and assured she was inbound from Rome on the next available flight. Of course, it had to be one of the Aussie contract crew because mainline Alitalia people refused to work with us. We boarded the passengers and waited. Sure enough, another Alitalia airliner parked nearby, and the ground handling people shuttled her over to us. I met Kerry at the door as she came onboard. She was a little breathless, but asked,

"Oh captain, will my bag make it in time for our flight today?" I replied that I wasn't sure, but not to worry as she turned to join the rest of the crew.

"Poor Kerry will have no clothes to wear in New York for three days." Lamented the Senior flight attendant, starting the running joke that lasted for the rest of the trip. We all took turns in teasing her about having to wear her underwear inside out etc. However, she shut down our incessant teasing three days later as we were taxiing out for departure back to Milan. As normal, the long line of aeroplanes snaked halfway round the airport heading slowly for the runway. Kerry entered the flightdeck, she'd come all the way from the rear galley.

"Just to let you know boys. I DID manage to buy some NEW UNDERWEAR in New York – LOOK!" she said as she lifted her uniform dress right up for us to admire her knickers and stockings. Then with a 360-degree twirl she bounced out into the cabin again. It worked - we were speechless.

In Milan, a lovely Irish pilot used to regularly host big lamb dinners that were very popular with the cabin crew. Other pilots enjoyed a real homely atmosphere, with their families living with them in the Milano Tres apartments. This was a development built by one of Silvio Berlusconi's firms no less. We always knew that the contract would have a limited life span. In some ways we were a victim of our own success. According to leaked info from the finance department of the airline, our Boeing 767s were flying passengers at approximately 40 percent, per seat of the cost of the mainline jumbos. So, our little fleet was making money for the company – something rare and unheard of for Italy's national carrier.

There was still the problem of a few hundred DC-9 pilots sitting at home on full pay. Gradually they returned to work. They were led at first by the training captains who came with us and sat on the jumpseat in the cockpit while they learned the operation of the B767-300ER and then the rest of the line crews came too. We were a little sceptical that all the DC-9 pilots would be able to handle the larger jet – it was a big ask really.

Our scepticism proved correct when one of the all-Italian flightcrews managed to severely damage one of those beautiful 76's in a landing accident in New York. It occurred after flying the Canarsie Approach for Runway 13, which is a particularly challenging arrival. In fact, I would rate it more difficult than the Kai Tak checkerboard approach in Hong Kong and I have flown both. The aircraft remained in a hangar at JFK for months while the repair team from Boeing rebuilt the landing gear and sorted out the hull.

But that all happened after I had left the contract. I was invited to join Royal Brunei Airlines (RBA) out in Asia on the 767 fleet. It was too much of an attraction to miss and I had promised my family a big adventure in the Far East. My bags were packed, and I left Milan on a mainline Alitalia Airbus A321 from Linate airport just after arriving from my last flight to the States.

I was more than a little weary as I climbed the stairs in uniform, carrying my hat and flightbag, but I felt much better when I was greeted by the beautiful smile from the Senior Cabin Crew member at the door. "Buon Giorno Commandante! We upgraded you to business class!" The crew then made a bit of a fuss of me, bringing newspapers, juice etc. before take-off. As I took off my jacket and relaxed in my seat, I knew I was going to miss this, The Italian Job.

*BACKSTORY: Anyone who was on the Nordstress contract in Italy knows what a good time we all had. You could say it's been all downhill since then. Halcyon days indeed. *Alitalia had a fleet of more than 30 DC-9s in the early '90s. When they were grounded in '96 all of the pilots stayed at home. **There is some doubt as to whether Neil Armstrong transmitted the words, "Good luck Mister Gorski" and in fact he always denied it in interviews. However, you have to ask the question, "If he didn't say it, who would invent something like that and why?" and his family really did have the Gorski's as next door neighbours in Ohio when he as a kid ...*

19 IT'S NOT ABOUT THE MONEY

"How do you know when you've got an airline pilot at a dinner party?"

"Don't worry he'll tell you!" So goes the old joke which incorrectly implies that all airline pilots are braggards, male and overconfident egotists. These days the reverse is true. Our profession is not what it used to be. There was a time when strangers might have been suitably impressed, but now their reaction is a mixture of pity and diffidence. They know our industry is on 'life-support' and those of us still working for the airlines are doing so in much reduced circumstances.

So, "what makes our occupation desirable even now to the younger generation?" - In the glory days, expat pilots working in the Middle East could rely on their pot of gold at the end of a 15-year stint. They could retire early on their return to Europe. At the legacy carriers of the EU, the situation was similar. At a youthful 55, a well-remunerated mainline captain would close the logbook and with generous pension provision, would hardly notice a difference in their income. Sometimes it would be MORE than they were earning flying the line! In Greece, the pilots of the great Olympic Airways used to say they could afford to buy a new car every month.

One of the advantages of having a long career is that you get to see a lot of changes, not all for the better. Certainly, the return on investment for new joiners is less sure than it used to be. Yet there is still a queue waiting to handover £100,000 plus for initial training and type rating. That training takes 12 months plus, in most countries without any monetary support from the airline or the taxpayer. Even if the newbie aviator lands the job of their dreams; their salary expectations are bleak. Some carriers even operate 'pay to fly' schemes – where co-pilots pay their airline for the privilege of sitting in the right-hand seat.

All things considered; it cannot be the money that attracts us. My own story is like many others. I always wanted to fly, a passion from before the age of 10. Am I disappointed with my choice of career? Not a bit of it. Is there anything I would change if I had to do it all again? Nothing. You see, it is still a bit special. Not everyone can do what we do, granted, some wouldn't want to, but there are many more who would love to. For whatever reason, it did not work out for them, although they wished it had. Those who fly for a living or who have done so are a little different. It is an odd choice of occupation and although not as glamorous as it was, it remains fascinating. So much so, that after a few decades you find ways of avoiding telling people what you do for a living. It is not elitism, trust me. It is just that I find it more interesting to learn about what strangers do for work, preferring to listen than to transmit.

Which is why, now when a stranger asks about my work, I often tell them "I'm in aluminium exports," which usually results in a lack of further questions. But if they ask more, I answer, "Oh, *aluminium tubing* mostly... but tell me about what YOU do?" and this then diverts the conversation.

A yacht called Crosswind – the clue is in the name

A couple of years ago I met a stranger in a boatyard in Greece. He was a long way from home. An affable Australian with an easy manner and happy to chat, waiting for his immaculate sailboat to be launched. Obviously retired, he was late 60's and dressed in scruffy shorts and raggedy tee-shirt. Clearly an experienced sailor, it was a delight to talk to him about his travels in the Mediterranean and beyond.

Not only that, but something else, he had a certain calmness... I couldn't quite put my finger on it. There came an appropriate moment and I said, "You're clearly enjoying retirement Steve, but what was your occupation?" He looked at me, smiled and said easily, "Ah well, I was moving equipment around... y'know..." as if he'd had the most boring job of all. Then we moved back to boats and sailing, a mutual interest. A while later, he asked me what I did, to which I grinned and said, "Well, I move equipment around... y'know..." He laughed, *"Yer a pilot aren't ya?!"*

It may not be the best paid profession anymore but, for some of us it is still the best job in the world.

✈

BACKSTORY: I still smile when I think about how I happened to write this piece. It is shorter than my usual articles because Dominic Perry, Deputy Editor at Flight International Magazine, stipulated "no more than 800 words."

I did my best. I edited and re-edited the story in August 2020, aiming for precisely 800. Not only that, but he also explained in his first email almost apologetically "We can't afford to pay for these contributions" which I thought was funny. So I titled the piece, "It's not about the Money."

✈

20 LEARN AND LIVE

The first officer called, "Eighty knots... Vee One... rotate!" and the captain started to raise the nose of the airliner. The pitch attitude began to increase. Then without warning, there was a huge noise from the left-hand side and the left engine fire warnings all illuminated. This was an experienced crew. They barely exchanged glances as the captain continued the takeoff.

As the severely damaged engine began to fail, he used more power on the live engine, controlling the flight path and correcting for asymmetric thrust. "Positive climb!" Now the first officer's voice was a little strained, but the captain replied calmly enough,

"Gear up." The first officer moved the landing-gear handle to the up position and the undercarriage made its usual grinding noise as the wheels retracted.

The captain relaxed just a little bit. He was calm, he was in control, flying well and now that the drag was away, he knew the aircraft would accelerate and climb. The fire bell now started after its inhibited period and the first officer cancelled it immediately to prevent the distraction. At 400ft the captain called, "Confirm the failure." His voice was authoritative... and reassuring. He had practised this procedure many times before in a Simulator and was well in control of the aircraft, and his first officer was an experienced pilot with several thousand hours on type.

"Left fire handle, Confirm...? Pull and rotate..."

"We have an engine fire on the LEFT side!"

The first officer had regained his composure after the shock of seeing all the warning signals and knew what he must do next.

"Roger that. Carry out the memory items for a LEFT engine fire," responded the captain as he trimmed the aircraft to obtain the best climb performance. Just after takeoff and at high gross weight, the aircraft was struggling to perform well and it would be a shallow climbout. The first officer did as he was asked and called aloud each of the actions as he was about to take them so the captain could monitor what he was doing.

"Auto-throttle, Off. Left thrust lever, Confirm? Closed. Left fuel control switch, Confirm? Cut off. Left fire handle, Confirm? Pull and rotate. Firing the first bottle. Starting the stopwatch."

Thirty seconds is a long time when you are watching an engine fire warning to see if the extinguisher will do its job.

"My radio!" said the captain as it was happening, "I'll put out the Mayday" "Mayday, Mayday, Mayday, this is..." His radio call was textbook and completely clear.

"Call approach frequency one, one, niner, decimal four when able, they have you on radar." Advised the controller in his immediate response.

"Firing the second bottle!" Now the first officer did look concerned as the engine fire warnings remained illuminated and the first extinguisher had not put out the fire. The stopwatch made a loud click in the flight deck as he thumbed the button. Another 30 seconds passed and the warnings remained. Now it was serious and he knew it.

"It's still on fire!" The first officer practically shouted across the cockpit.

The EICAS warning confirmed LEFT engine fire

Now a new tension filled the cockpit. Both extinguishers had been used and there was still a fire. The commander's professional composure gave way.

"S***! Tell them we want a priority approach to land. Do it now. We need to get this thing on the ground bloody quick!"

The Captain put the autopilot in and started a turn back towards the airport. There would be little time to warn the cabin, he thought to himself, *'I'll just have to call Brace, Brace on final approach... if we make it that far.'* It was seven minutes since rotation and the incident began. They knew that every second would count. How long before the wing spar burned through, or the engine exploded? They could only guess, but with an uncontained engine fire and no way of extinguishing it, they were likely to find out.

The first officer was now busy on the radio to air traffic control... and calling out the speeds for landing... and reading through the rest of the drills from the emergency checklist. As a high-speed, 'one-armed paperhanger' he was doing quite a good job.

The captain glanced at him, looking for the right moment to give an abbreviated, emergency landing brief – then he looked down at the engine fire handle on the centre console. It was still illuminated bright red, but there was something wrong. Even more wrong than the fire warning. The captain let out a very unprofessional oath as realisation dawned. The extinguisher bottle discharged lights were not lit! The fire extinguishers had not been fired! He reached down and twisted the fire handle first to one side and then to the other until both 'bottle discharged' lights illuminated.

Simultaneously the fire warning went out... and then I hit the 'Total Freeze' button on the instructor console in the back of the flight Simulator. Everything fell silent. Being an experienced crew, they virtually debriefed themselves, with only some gentle facilitation from me as the examiner, to ensure that they understood that it was essentially a breakdown in Crew Resource Management (CRM) that led to the technical failure of not actually firing the extinguishers.

Although the right-hand seat pilot turned the fire handle, he did not 'hold against the stop for one second' as specifically instructed in the Emergency Checklist. With full situation awareness of the extinguishing system, he would have been looking to see that the bottle discharge lights illuminated confirming his actions. The captain took at face value the statement by the first officer, "It's still on fire!" to mean that although both extinguishers had been used correctly, they had failed to put out the fire. The old adage, 'Never assume, check' was repeated in the debrief along with a review of the decision-making process to make certain that they were aware that if they had employed this correctly as a crew, the omission would have been spotted earlier.

As examiners of professional pilots working in Simulators, we are often viewed as 'Trappers'. Some of our colleagues see us as being there to impede their progress and only to point out the human failings which we can all make in stressful situations. Nothing could be further from the truth. However, there are times while observing when we must stay silent and resist the temptation to intervene in order that the critical learning process is not interrupted. The above scenario actually occurred. While it was unfolding, I had a very real urge to warn them and reassure them that I had not been the b*****d examiner who had given them an engine fire which they could not extinguish. As I pointed out later, where would the training value have been in that?

Our job in the examining role is to keep it as real as possible in the Simulator, so that not only can we make a true evaluation of pilots' performance, but more importantly ensure that the learning points have maximum emphasis. Many is the time that we try to get this across by explaining to the pilots that we know what it is like to be in their shoes and that we are crucially aware that it is "dead easy to fly from the jump seat." In a similar way, instructors and examiners in General Aviation will make every effort to maximise the training value such that their trainees and candidates realise their own potential at every stage. A good example of this is when flying instructors progressively increase the workload on their trainees as they approach the time of their first solo flight.

While the following examples may not be textbook instructing practice, they do make points about how one might help a student to learn, either in the Simulator or in a real aeroplane, by taking a student safely out of their personal safety zone and developing their abilities. To reassure one particular student who was nervous about the idea of flying solo and carrying out the landing alone, the instructor deliberately took his coat from the back seat of the aeroplane and put it over his head so he couldn't see.

"Talk me through this circuit, but I'm not going to watch." The student did just that, proceeded to land the aircraft without any problem and went solo shortly afterwards. As I say, not standard instructing practice and probably not applicable to every student, but it did build confidence and broke the student's reliance on the instructor.

An extreme example of an instructor making sure that his student learned the lesson 'always keep your hand on the throttle after takeoff' occurred many years ago when I was going through basic flying training. The aircraft was a de Havilland Chipmunk which had a tandem cockpit and the instructor sat in the rear, with full dual controls. This particular instructor was keen to see that his trainees learned their lesson well and if the hapless student in the front seat forgot to keep his hand on the throttle lever, the engine suddenly lost power and revs as the throttle was smartly closed in the rear cockpit. Immediately the student put the nose down to glide the aircraft and prevent a stall. He tried to open the throttle, but it was as if it were jammed shut.

Then came the growling voice in his ears from behind.

"Now y'see lad what 'appens if yer take yer hand away for even a moment just after takeoff. Th'engine dies and yer throttle jams shut and we're goin' t'have ter force land th'aeroplane!"

Again, maybe not an instructional ploy you'd condone in every situation as it raises safety issues in itself, but it certainly taught the student a bigger safety lesson. Fortunately the days of training young pilots by scaring them half to death have gone. Trainees will always be challenged to give of their best and to achieve their true potential by good instructors who make the effort to push them further.

This is especially true of students who are passing with ease, whichever course they are on. In basic flying training, for instance, it is a simple matter to convince the apprentice navigator that they have perhaps put 'the drift' on the wrong side as they are conducting their navigation exercise. A mental problem which is so easy to solve on the ground all of a sudden becomes a worrying mental challenge when you have half-a-ton of flying metal in one hand doing 100mph, and you are running for your first turning point.

"Are you quite sure the drift is on the correct side for the wind?" is normally all that is required to prompt a hurried replan of the calculations with the inevitable errors, which are only discovered on the next leg of the navigation exercise as the aircraft goes completely off-track.

We all make mistakes from time to time. So long as the instructor obtains the training value out of the lesson to ensure the trainee resists the distractions that can cause errors and crosschecks their work every time, then the exercise is a valuable one. In his iconic book about commercial flying, 'Fate is the Hunter', Ernest K Gann tells the story about a hoary old captain who used to challenge young co-pilots by lighting matches and holding them under their nose while they were trying to fly a complicated approach on instruments in bad weather.

Each time the sulphurous, smoky burning substance came near their nostrils, they had to blow it out, while – eyes streaming – they also had to fly the airliner on instruments. Not surprisingly, Gann found that his own bad weather instrument flying improved markedly when he was flying with other captains who did not try to distract him in such a manner. Was that good instructing or not? Was the end worth the means? It may be that it's down to the particular co-pilot and the individual situation. But it all goes to prove: if you simulate circumstances at least as bad as you can expect in real life, it will have a beneficial training value, and will increase your safety immeasurably.

BACKSTORY: Originally published in the UK CAA Flight Safety magazine "Clued Up" back in 2008. The B767 simulator story was real. The editor added a couple of bits about General Aviation flying training to my words to make it relevant to GA pilots as well. They preach to a large congregation those people at the CAA. Anyway, I was not unhappy with the overall result and if it made one or two pilots think twice, that would be good.

Boeing 767 Simulator interior – as real as it gets

21 SHARPENING THE RUSTY BLADE

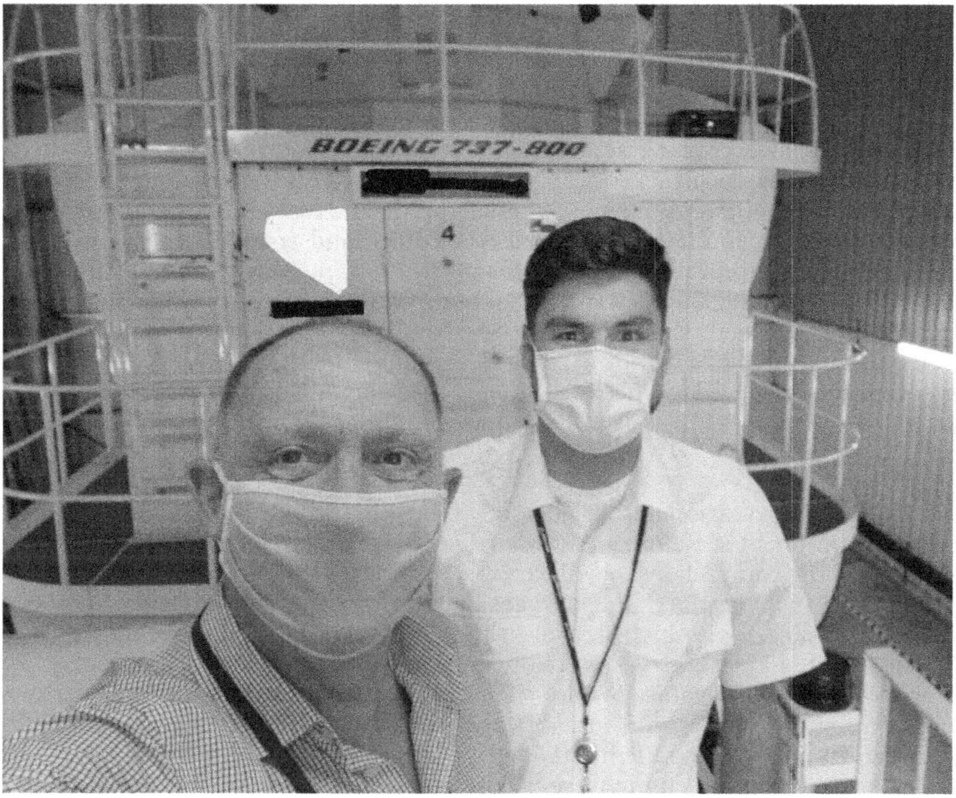

"Mate! Just remember, we're a team here and do not hesitate..."

After a full four months of not flying the line, it is time to become current again. This means Recurrent Simulator Training in the full motion Simulator. Not just training but checking too. Every six months airline pilots are required to undergo an Operator Proficiency Check and every 12 months a Licence Proficiency Check. There's no way round it, as the name implies, these are pass or fail situations. Flying Commercial Air Transport aeroplanes is a serious business and the airlines take their responsibility diligently to ensure their pilots are up to scratch. Not only that, but from the subjective point of view, we professional aviators are acutely aware of the huge numbers of our colleagues who are unemployed now – there is added pressure on us to make certain our performance is more than acceptable. If we fail, there are plenty who would take our place.

It is hard to convey the intensity and level of focus required during recurrent airline training. We are sitting in the briefing room (mask clad and socially distanced in this time of COVID) for a full 90-minute briefing prior to entering 'The Box'. When I say 'full', it really is. There are barely any spare minutes if we are to cover the required syllabus of briefing items. I estimate we are just over halfway through and have already discussed some CRM scenarios in detail – using actual reports of previous air crash investigations.

Also, we have covered several B737 technical systems in depth. The training captain asks plenty of oral questions to make sure we know our stuff. Numbers, numbers, numbers... pressures, distances, volumes, speeds, angles, altitudes, limitations, they all come thick and fast. You must stay focused to keep up. And remember this is after four months away from the operation. It is a challenge. In a moment of respite, I liken it to having a rusty knife blade that needs sharpening. We two candidates/trainees (call us what you will) are the blades and the training captain is the grindstone. Well he has got that old grinding wheel up to speed today and we are getting sharper by the minute.

✈

The briefing session marks just the final phase in the pre-simulator preparation process. Before we even arrive at the training centre, we have completed the comprehensive E-Learning Technical, CRM and SOPs course online. This includes the exam packages that are again pass or fail items. The pass mark is set quite high, but even then, we all aim for 100 percent if possible – it is kind of expected really. You have to say, excellence doesn't come cheap in this game. While a lot of the information we process in preparation to fly the sim is already familiar to us, there is much that has changed due to new pandemic-related procedures. Sometimes these new procedures align with common sense, however others are counterintuitive and require our full concentration to carry them out.

Finally, the briefing is completed, and we are ready to launch. We head out the door towards the Simulator. *My God this feels like a moonshot,* I think as we approach the large white box-shaped machine standing on its hydraulic legs. We sanitise our hands and leave all liquids on this side of the drawbridge, as the training captain gives us a short safety brief. Of course, we have been in many full-motion simulators before, but not all have the same emergency egress techniques and facilities.

We pay attention; our lives really might depend upon it. There is a lighthearted moment when the trainer points out that the very real looking portable fire extinguisher inside the box "is a dummy", (Simulating the one we carry in the flightdeck at work) while another, much larger one, at the back of the Sim is the real McCoy. I note that it is a big volume CO2 extinguisher, which would be very useful if we had to fight our way out of the building. Our tutor also indicates the rope ladder that we would have to throw from the side of the outside balcony to descend to the ground in an emergency, possibly in complete darkness.

Now we are seated, strapped in the Sim and the familiar feeling comes back from all the instruments, controls, levers and screens. Almost by magic, my automatic locomotor co-ordination takes over and the subconscious part of the brain guides my hand and arm movements to the correct buttons and switches. I think this is known as muscle memory.

This is the bit which is 'like riding a bike'. There are not many dials anymore of course as it is pretty much a state-of-the-art 'glass cockpit' airliner. Even the analogue display gauges on the screens are computer generated. Now it is time to get the machine setup and ready to fly. Naturally the weather is awful – that often happens in the Simulator I note with a wry smile. Outside the building I know it is one of those quite rare English summer days with sunshine and blue sky, but I try to put that thought out of my mind. Yes, I would rather be anywhere else right now. The background electrical humming of the Sim takes on a different sound as the instructor announces solemnly,

"Motion coming on gents!" having first ensured that we are strapped into our seats. There is a slight lurch as the hydraulic jacks take up the strain and the whole flightdeck and instructor station is lifted a few metres into the air.

Tension rises in the cockpit now and I look across at my younger and much less experienced colleague. This is only his second recurrent Sim session after his initial Line Training on type. He has very few flying hours and I know that he is looking to me for the right sort of guidance and leadership to get us through the next five hours. Yes, no typo there, the full-flight Sim session is planned at five hours long. There will be a break for tea/coffee halfway through, a long time from now. Here and now, right at the start, is the time for me to set the tone for our CRM interaction. I appreciate that the guy in the other seat will probably feel there is a steep seniority gradient between us, and I am anxious to flatten that as much as possible and quickly too. I say to him,

"Mate! Just remember, we're a team here and do not hesitate to tell me if there is anything I do or say which is at odds with the way you see things. I am open to any suggestions which you have to make things easier for us." I can tell he is pleased that I have said this, especially starting with the old back slapping term "Mate" It is important that he practices good monitoring skills while I am Pilot Flying (PF) because much of my attention will be consumed with accuracy. There are tightly prescribed limitations on the accuracy of our flying, even as we face a shed load of distractions from emergency situations.

Then we are off. At last, the aircraft is rumbling down the runway, both engines spooling up towards takeoff power. The sights, sounds, feelings and movements all so real. We play our part in this performance too, trying as best as possible to adhere to our script, I hear "Vee One... Rotate!" and I increase the back pressure on the control column to unstick first the nosegear, then the mainwheels from the runway. There is a tendency with the B737 (just after the maingear has got airborne) for the pitch attitude to 'hang' at 10 degrees above the horizon and I increase the back pressure slightly in an effort to maintain a steady rotation rate of about 2.5 deg/sec. I call "Gear UP!" and he responds, "Positive Rate" and selects the landing gear lever to the UP position.

As I see all the landing gear lights extinguish out of the corner of my eye, I smile – we are flying now for sure. I know that the landing gear and doors are all retracted and breathe a very small sigh of relief, one less thing to deal with. Then the flaps are all retracted shortly after this and the after-takeoff checklist is completed, but no time to relax. Once the autopilot is engaged and crosschecked, I hand over control to him and start to set up and brief for a non-precision approach back into the airport we have just left.

I run through the setting up of the instruments and navigation aids with reference to the approach plates on the iPad. Simultaneously it seems I am setting up the Flight Management Computer routing so we can make best use of the automatics. When that is done, I itemise the threats which I can see to our approach and ask him if he can see anything that I might have forgotten?

He responds with reminding me that I have not set up the Fix range ring circles in the FMC so that they are shown pictorially on the Navigation Display (ND) screen. Damn! I silently curse, but he's correct and thank heavens he spotted that, because now is the time to set them up, not after we have started down the approach! I thank him genuinely for this intervention although I know that the trainer behind us is noting down my omission. Rusty after four months at home? Yes, I should say so!

I now rattle through the standard briefing to ensure we're going to fly the correct profile and cover all the contingencies too. Even at the end of my speech, I ask him again if there is anything else, but no, this time I've nailed it. Then he hands over control back to me. We run through the descent and approach checklists and I ask him to let ATC know we are "Ready to make the approach." The controller (aka training captain playing the part of the ATC radar controller) clears us for a straight-in approach with too few track miles. My colleague repeats back the clearance on the radio then looks across at me questioningly. He senses all is not right. I grin broadly and ask him to,

"TELL Air Traffic we need more track miles please!" If we had simply taken their radar vector 'straight in' there would have been a High Energy Approach (HEA) situation and that would have led to a Go-around from being unstable on finals. *That ain't gonna happen today!*

As we progress through the five hours together, with approach after approach and takeoff after takeoff, I can feel us getting back into the groove. It's all coming together and despite all the fires, catastrophic failures and other less dramatic emergencies, we work as a good team. At no time is the safe outcome of any scenario in doubt.

Not perfect of course, but we get sharper and sharper as we approach the end of our time in the box. Finally, the trainer switches off the motion. We both feel drained. The thought hits me, *tomorrow we must do it all again* but this time under examination conditions – the proficiency check! Well, I shrug my shoulders, we are ready. Bring it on, these two blades have sharpened up nicely.

BACKSTORY: No pilot really looks forward to their 6 monthly Proficiency Check. This time it was even worse, because we were all out of flying practice thanks to COVID. This was July 2020 and I had flown very little since March. We managed a good performance though and by the end of Day 2 we were killing it!

I wrote this piece because the experience reminded me how tough it really is to produce first rate flying standards under extreme pressure. We must process information so quickly. Plus we need to have outstanding situation awareness and anticipation. It's a bit like trying to drink from a firehose!

Some moments are impossible to simulate – you have to be there

22 CARRIER LANDINGS

On final approach to RW25 at Palermo, Sicily

I was born too late. They said that about my father too so perhaps it runs in the family. For different reasons. In his case he had all the manners and gentility of an Edwardian gentleman* – he even wore a cravat. Imagine that, a cravat in the 1960s! I don't know anyone else who wore a cravat back then well, apart from Mick Jagger. You see, now ***there*** was a man, born out of his time... Louis McBride, not Mick.

In a similar way, when I joined the Royal Navy Fleet Air Arm in the early 80s to fly fixed wing and train as a 'Stovie', the proper aircraft carriers had gone. The RN was down to just three light aircraft carriers, the glory days of the large Audacious Class ships were finished. All those Phantoms and Buccaneers; scrapped and no angled-deck operations anymore. But, years later I found myself flying operations with some similarities. The challenge of landing (and taking-off) from short runways is something a little bit special in the airlines.

Off the top of my head I can think of 8 airports that pose significant challenges to jet airliner operations, mainly due to performance considerations. In no particular order, they are Corfu, Palermo, Calvi, Funchal, Punta Delgada, Kos, Skiathos and Mykonos.

No doubt many others could easily be added to the naughty corner, but I think you would find these are the main bad boys. Most European airline pilots will have their own personal story about how things went wrong on a visit to at least one of the above airports. Bear in mind that these locations are all considered more difficult than usual and that's assuming GOOD WEATHER. Some of them are classified as Category C airports. There are only three categories and C is the most potentially hazardous. Let's look at why.

One example is Heraklion Airport on the island of Crete. When you come into land on Runway 27, the actual start of the concrete strip appears to be located exactly at the top of a cliff. Seriously, the sight is more than a little disconcerting when you see this for real on final approach and even more so at night! The first time I ever landed** at Heraklion (or 'Eric-the-Lion' as we used to call it) was in a Boeing 757 from Manchester back in 1989 during my initial line training for the airline. It was a night-flight and my Line Training Captain (LTC) said it would be good experience for me to operate the outbound sector of this night flight.

As part of the briefing for the descent and approach to land, he informed me that I would probably see the cliffs in the landing lights on short finals – he was right. He warned me against being distracted by this. While I appreciate it is only psychological, it is a major distraction to be cognisant of the fact that land is so precious, they have taken up as much space as possible to build 'thunder road'.

If departing from Runway 09, (the other end) you are critically aware that this runway *ENDS at the top of the cliffs...* there is nil overrun area should things go wrong. You then have a new respect for the meaning and use of the call **"Vee-One"** during the take-off roll. Remember the definition of V1, is that the go/no-go decision should have already been taken by the time that call is **complete**. A late decision to abort the take-off has been the cause of many take-off accidents. It reminds me of the words from Boeing in our Airplane Flight Manual; 'Stopping on the paved area is **NOT ASSURED** in the event of a 'STOP' call made **AFTER V1'**. If you do this at Heraklion on RW09... you will end up in the sea. An unpleasant thought.

Funnily enough, even though Heraklion airport left an indelible memory for me, it is not a Category C airport for most airlines – it is classified as 'Cat B with Restrictions' by the operator. These restrictions for any of the B and C airports may be captain's only landing/take-off, Simulator Training required, Special Briefing to be signed for as having been given etc. All these steps are taken for one reason - to mitigate the threats posed by these rather special locations. Another airport which is 'only Category B...' is Corfu. However, in common with myself, many commercial pilots have a story or two to tell about this one.

There is one runway laid out north/south. It is designated RW34 and RW16. Now I can tell you that if the airport is operating on RW34, there are fewer issues straight away. For one thing, there are instrument approaches that nearly line up with the runway centreline. Not an ILS... oh no, that would be too easy – there are no 'precision approaches'.

In fact, the reason that the final approach tracks are offset is due to the high terrain at 5 nautical miles on final approach. Well it's not actually ON the final approach, it's JUST to the west of the extended centreline for the runway, but the thing is this... At 5nm on final approach with a normal 3-degree profile (i.e. 300 feet/nm) the airliner should be at 1500 feet. The height of the terrain? ... 1490...! You can see the problem eh? Which is why many operators designate RW34 operations at night as ***Captain's Only landing.***

Again, it's a 'carrier-deck landing' this one, as the threshold for RW34 starts in the lagoon! The Royal Grand Hotel is just to the east of the threshold which has long attracted spotters and keen aviation photographers. Such observers will tell you, this runway can provide some interesting sights. That said, from the flightdeck viewpoint, provided you take into account, short runway, rough surface, high instrument approach minima, offset finals, maximum flap for landing, normal reverse thrust (be ready to select maximum...) ***and*** a high autobrake setting, then you'll be fine.

The B737-800 autobrakes are usually set for 'Max Auto', because there is quite possibly a tailwind for landing... oh sorry, didn't I mention that one? And of course the airport management prefer to use RW34 due to noise. They will often continue to use it for landing long after we (the pilots) would have preferred RW16 due to the local wind conditions. But RW16 can be even more of a challenge frankly.

You see to land on RW16, there is no published instrument approach, so you must use a VISUAL Circle to Land procedure after first carrying out some sort of let-down through cloud. I distinctly recall one of my first night arrivals at Corfu shortly after being promoted to Captain on the B757. The weather forecast was pretty foul, so we were carrying a lot of extra fuel as we expected to maybe hold for some time prior to making our approach if there was an active thunderstorm overhead the field when we arrived.

Sure enough, there were active thunderstorms in the vicinity as we descended with the radar controller, but the worst news was that they were using RW16 (in those years it was RW17). Ouch! We therefore briefed for the break cloud procedure from the Garitsa VOR, prior to making our visual circuit downwind LEFT-HAND for the runway. The downwind leg for the visual circuit can only be flown to the east of the runway, due to terrain to the west. But you can't fly very far to the east or you run the danger of infringing Albanian airspace which in those days was more of a worry than it is now.

On final approach to Corfu RW34 – another aircraft carrier

Anyway, we were flying downwind at 1500 feet above the airfield (elevation is sea level), with landing gear down and Flap 15 in the 757. A great configuration for slow manoeuvring. I was flying with the autopilot and autothrottle engaged still (using Heading Select and Altitude Hold modes at this point). This gave me extra situational awareness and therefore spare mental capacity, which was certainly needed on a night like this. Thankfully, because I was still in my first couple of months in the left-seat, I was teamed up with an experienced First Officer, but of course he had never landed on Corfu's famous RW17 either. The night was as black as your hat and now it started to rain, ewww... I think we passed underneath some cumulonimbus because it was very turbulent as you would expect.

I was still confident that we were visual with the runway/airport out on my left-hand side and thank God for the moving map display, but now it started to rain. I called "Wipers On" as I was still concentrating on visually tracking the airfield which now began to disappear behind the hill in between us. I looked forward through the front windscreen with horror to see that the rain contained SAND! Aaarghh! *That wasn't in the plan AT ALL!! Damn!* The front screen was a blurry mess of brown crap, being swished to and fro by the wiper blade. I couldn't see where we were going. I felt rising panic. If this persisted as we turned on to finals there was no way I could land the aircraft if I couldn't see where we were going. *We would have to Go-around... and THEN WHAT!!??*

Meanwhile, there was nothing else for it, but to continue to the point at which we had planned to turn base leg and start our curving descent onto final approach. Fortunately the rain eased off such that we could just about see what was happening out front. I cross-checked that the FO was happy with our position, disconnected the automatics as we completed the final turn and called for landing flap setting. Now all I had to deal with was the deceptive visual illusion of being LOW due to the high ground under our flightpath that slopes all the way to touchdown.

I locked onto those PAPIs, Two Reds, Two Whites and ***did not let go.*** As I recall, we did not float in the landing, I just stuck it down and made sure she was in the right place! It was so satisfying to feel the autobrakes kick in with their huge retardation forces, followed by the huge noise from the reversers. God bless the designers who produced an airliner with so much surplus braking capacity. Seriously, that aircraft with its big four-wheeled bogies on each main landing gear is a star performer!

As you may have gathered, frequent common denominators for Cat B and Cat C airports is the fact that they are built on islands. Land is at a premium and they use every square metre. You can also predict that the paved surface is going to get close to the beach. Oh! I forgot about Santorini - now THERE'S A STORY or two, but another time eh?

It is true that over the years the decks on my 'aircraft carriers' did not move, but short runways with complicated approaches have still provided me quite a challenge. All things are relative. Our flying machines are 60 tonnes and up with correspondingly high threshold speeds.

✈

BACKSTORY: I still have one of my dad's cravats, although I do not expect to wear it myself anytime soon. It makes me smile when I picture him wearing it when I was young.

**My father (Louis) was born in 1919 nine years after the end of the Edwardian Period (1901-1910).*

***I was 'Maxed-out' with information at that stage in my flying career. I only knew that Heraklion airfield was close to the coast, somewhere in Greece. It was at least a week later when I realised it was on an island called Crete!*

Corfu: In this part of the world, it is not unusual to have sand carried in thunderstorms. Straight from the Sahara – the southerly wind for landing was a clue which I missed.

23 THE AIR INDIA GIG

B767-300ER G-CEFG – taxiing out at Heathrow

My mother looked surprised when I said I was planning to go to Birmingham to get my visa to go to India. I knew there would likely be a big queue, so I wanted to arrive early.

"You won't need a visa for India darling", she said in her very poshest British accent. "WE OWN INDIA!"

"Mum, they kicked us out in '47 when they gained independence", I replied when I finally stopped laughing, "we don't have an empire anymore!"

Even now some 15 years later, I am still not sure if she was joking. That was the thing with our mum, she was smart. She kept us all guessing and she had a great sense of humour. Her story about how she was "responsible for starting the Second World War" was typical.

As a nine-year-old Catholic girl being cared for by nuns, she was instructed to pray every night at bedtime that there would not be a war with Germany. The year was 1939 and at the end of August one night she forgot to pray. Imagine her horror the next morning when she learned that Britain had declared war on Germany? She felt so guilty she expected to be arrested by the police! Poor thing, it took her years to get over that one. But I digress.

Standing outside the Indian Consulate in the rain at 6am it is easy to kick yourself for not bringing an umbrella. They don't open till 9. Naturally you don't want to lose your place in the queue so you snuggle up against the brick wall of the building, pull your collar up and wait. Behind me most of the people waiting appeared to be Indian and I was at a loss to understand why. Then it dawned on me. They must all have British passports! So, they need visas to visit the sub-continent. Okay, now I don't feel so bad.

But let us wind the clock back to the time when I received a phone call from Jack Brown to offer me a job. He was the Head of Training for Flyglobespan, an airline headquartered in Scotland. It was 2006 and I was flying as a Boeing 757 training captain for Astraeus based in Gatwick. I enjoyed flying for Astraeus, nice people and interesting destinations. I had once told Jack that I might be willing to work for him, but I hadn't fully decided. Now when we spoke, he wanted an answer and he was at his most persuasive,

"Look James, you'll be based in Manchester as a TRE training captain and you'll ONLY fly to Cape Town or Toronto."

I had to admit it sounded like a cool job, but I was also looking to work part-time so asked if I could,

"Work two weeks on and two weeks off Jack?"

"Yes of course no problem James" Jack assured me, "you'll do your two weeks either in the Sim or on the line. Are you coming?"

Well, it was a dream job, so my answer had to be yes. I couldn't wait to tell the family; they would be so pleased to have Dad at home much of the time. *Cape Town and Toronto – two super destinations, wonderful.* I have to say it wasn't Jack's fault, when it all changed after I joined. He was not responsible for Flyglobespan signing a wet lease* contract with Air India to operate to and from London.

I did fly to Cape Town and Toronto a couple of times and very nice they were too, but then I got another call from the chief pilot. He told me about the Air India deployment and kindly asked me to help by operating on the routes. He was quite flattering and said the company needed "experienced captains that could be relied upon to get the job done." So naturally I said "Yes" (again).

I remember thinking, *at this rate, I will be volunteering for things soon.* In the armed forces the conventional wisdom is to never volunteer for anything but I don't believe in that philosophy. Sometimes good things happen to those who volunteer. So it was with that positive frame of mind that I journeyed south to London Heathrow, with my visa in my passport. I mean, okay it was not the job that I signed up for but it would be interesting. I had never been to India.

The country held great fascination for me. As a schoolkid growing up in 1960s Britain, I was aware that my nation used to have an empire. Even in those days our school atlas books had maps which showed the British controlled overseas territories coloured in pink. There was a lot of pink. In 1901 the British Empire was the largest the world had ever seen and Queen Victoria was Head of State for nearly a quarter of the global population. India, the famed 'Jewel in the Crown', made up a large part of the landmass governed by London.

So the historical links between the UK and the Indian sub-continent were well established. To put the following into context, remember that the first ever powered flight by the Wright brothers was only in 1903. Air India, the national airline is one of the oldest in the industry. It was founded in 1932 by the visionary tycoon J.R.D. Tata.

He was India's first licensed pilot (1929) and flew the first commercial service in October 1932. That was a plane carrying mail from Karachi to Bombay (Mumbai), then on to Madras (Chennai). Tata Air Services became Air India in 1946. Juhu Airfield in Bombay (Mumbai) was the first civil airport in the whole of India. Perhaps it was no coincidence that the hugely wealthy Tata family owned much of the Juhu area of the city, right down to the beach.

My first arrival in Mumbai was as the operating captain on the Boeing 767 we were flying. I have to say it was all quite exciting and that comes from a pilot who has flown to lots of exotic destinations. This was the country I had learned about as a child in geography class. If I had any doubts about the challenge of maintaining the infrastructure at Mumbai's Chhatrapati Shivaji International airport, they were dispelled as we parked on Stand 51. The latitude and longitude coordinates sign painted on the wall was crumbling away and it looked a bit sad.

The crumbling infrastructure of Mumbai International

Not only that, but then I noticed the bamboo scaffolding on the side of the terminal building with a guy painting the walls. His lifeline was a piece of raggedy rope tied around his waist – yikes! If he fell off, that would likely cut him in two. Not good.

Health and Safety in Mumbai – still to be invented

I looked across to the south side of the airport where there was a very long line of pristine private jets. Close by were some old Boeing 747s that were clearly being cannibalised to keep the current Air India jumbo fleet running. This stark contrast runs through India like the Ganges. It is an inseparable part of the sub-continent culture. The tourist board even capitalise on it with their **'Incredible India'** advertising slogan. It's true, the country is totally incredible!

You do not realise how much though until you arrive there and see it all for yourself. For example, you arrive at an international airport in uniform as airline crew - not forgetting you are operating for Air India. Then you spend a ***very*** long time waiting for immigration clearance. Yes, you're in uniform and yes you have a valid visa in your passport and yes you have prefilled all the required documentation pertaining to your flight. And it still takes half a lifetime. Welcome to the Indian way of doing things. Of course they themselves blame the British for all the complicated bureaucratic processes. Well the Brits may have left them with bureaucracy, but we never intended they should make it their national sport!

After the frustration of obtaining access to the country, we were transported in a crewbus to the nearby Hotel Leela, part of the Kempinski group of five-star hotels. The building and grounds are beyond a set of large gates. There is security and a huge perimeter wall. We were still grumbling as we descended from the bus, only to be greeted by lovely welcoming smiles from the spectacularly costumed doormen. In fact, we should call them by proper title, these were commissionaires.

Hotel Leela Commissionaires (& my son Lawrence)

Everybody is ready to help with the luggage. All we had to do was walk through the doors into the beautiful hotel reception area. Everything was bliss. We moved from the smoky, humid heat of the outdoors to the fragrant air-conditioned, marble floored lobby. Enroute to the desk we were greeted by young women in Indian dress who greeted us with the traditional "Namaste." They very elegantly put their palms together in front of them as if praying with their elbows out at the side and bowed their heads. I felt a bit embarrassed not knowing how to respond, except to offer a stilted, "Good afternoon." I asked the first officer, who had been to Mumbai a few times before,

"What does Namaste mean?" His answer surprised me,

"It means literally '*I worship the divine in you*', but these days they use it more like a formal greeting. It's Hindi." I replied,

I love experiencing a foreign culture first-hand and to observe how people show respect for each other. *Hmm,* I thought, *Nah-Mas-Tay... Nice.* I would not forget.

At last we had arrived in Mumbai and soon we were flying two B767s on behalf of the national airline. Probably like me, you would be wondering why? What precise set of circumstances could have possibly led to this? The answer was simple. Air India had ordered a new fleet of 'Triples' (777s) from Boeing, but they had too few pilots with the B777 rating.

So, the airframes were being parked up without the trained crews to operate them. As a short-term measure, they leased in two of Flyglobespan's wide bodies. Initially the company sent Flyglobespan senior cabin crew to help the local Air India crew find their way around the aircraft. After a while this became unnecessary, which was when we became what they call a damp-lease. That is with aircraft, maintenance, insurance and flightdeck crew only.

The time zone difference between India and the UK took some getting used to. It is five and a half hours ahead of UK. Why the 30 minutes? The reason I was originally given that it was because Pakistan was five hours ahead of UK and there was no way that India would accept to be in the same time zone. However the real reason is because India is so large it spans two meridians. In principle the country could have two time zones, one hour apart. For simplicity however they chose to have one, that is 30 minutes different, but covers the whole country from east to west.

That crazy clock made me laugh whenever I saw it

The option to have just the one time zone is not unique. For example China, which could have five separate time zones (like continental USA), but instead they have just one zone based on Beijing time. It is eight hours ahead of UTC and covers the country.

This makes for some odd hours in people's lives. In Urumqi in the far west of the country for example, the sun doesn't rise until 10 am some days. Schools and train timetables operate at weird hours, national education examinations may take place in the middle of the night and restaurants stay open till the early hours of morning. All because the Communist Party wanted to have unified control of the population.

The oddness of the time zone was exemplified for me by the clock on the wall of the Air India operations room at the airport. It is the only clock I have ever seen with four hands. The red ones are UTC and the black ones are Local time. At first glance it is confusing but matches the Indian national psyche. In a well-meaning effort to make comprehension of the time difference easier they just made it more difficult! That clock used to make me laugh whenever I saw it. It has been up there for many years.

✈

G-CDPT – Your Palace in the Sky – who knew?

It was fortunate for Flyglobespan that the colour scheme of our company's aircraft was largely red and white. Similar in many ways to the Air India branded airliners flying for the national carrier. They must have saved a stack of money; all it took was a little more painting and we were looking like we belonged in Mumbai.

I was amused to see that we had been renamed "Your Palace in the Sky" on the outside of the fuselage (Who knew?) and that the windows had been given the fancy framing treatment all the way down the plane. Finally, the tailfin had been given the branding also, but some of the decals/paint did not last very long.

It was a measure of how frequently the machines went to and fro between the UK and India there was never enough downtime to get the paint refinished. Additionally, we were soon flying other international routes from Mumbai. We were certainly busy but not all the pilots had the same exposure to these frequent duty periods.

One of the main problems with India being so far away from the Edinburgh based crewing department was that it was not easy for them to keep control. Now and then the crew planners lost the plot as typified by what happened to Simon one of our captains, in Mumbai. He was about to report for duty one day, when he received a call from the crewing guys in the UK to say that he was to receive a roster change – somebody else would now be doing his flight to Heathrow tonight. Simon asked what he should do. As the plan was rather fluid... the controller told Simon,

"Look, just stay where you are and we'll call you back, okay!?" Needless to say, that call never came. Simon tried to call the crewing people a couple of times but the line was always engaged. So he did as he was told. Remaining as a resident of the wonderful Leela Hotel for the next seven days.

We were pleased to see him sitting in the lounge near reception reading the newspaper at his leisure while we all came and went in our uniforms. Later when we returned from our duties, we might see him in the bar having pre-dinner drinks with some other pilots. He was always in a fine mood it seemed and in no hurry to go back to flying the line with us. Eventually his mobile phone rang, it was Edinburgh.

"Simon...! Where ARE YOU?!" he replied cordially that he had been told to stay put in the Leela until crew control called him! Of course, they were exasperated by this and immediately found some flights for him to do. Much to our amusement, his mini vacation had come to an end.

On another occasion I recall there were about eight of us sitting outside by the pool, having a few beers in the warmth of the early evening. One of the pilots Dan, had called crewing to see if they could arrange a swap with another captain. The other captain's name was Dave. He explained what he wanted to do, which was to change a flight in a week's time, the guy on the other end of the phone was being very helpful and said,

"No probs, leave it with me and I will get it done for you both." As he closed the phone, Dan looked confident and relaxed that the seemingly efficient crew controller was going to solve the problem. No more than a minute later when one of the first officer's mobiles rang. It was the same crew controller who asked him,

"Any idea where captain Dave is at the moment, we want to get in touch with him?" So much for efficiency. One of the other pilots chirped up with,

"He's just taken off for London, Heathrow two hours ago!" To which we all fell about laughing. You would think that the crewing/operations people would know who was flying their aircraft. Clearly not always.

For the most part though we had a good life at the Leela. The staff looked after us magnificently. I even took my wife and young son back there some months later for a short vacation. My boy Lawrence was most impressed with the commissionaires. We flew in business class which was something we had not experienced since I worked for Royal Brunei Airlines (RBA) back in the late '90s.

✈

The cabin crew were lovely to work with at Air India. When we reported for duty at the ops control centre in the airport, the crew were all turned out very smartly in their uniforms. Some of the uniforms were based on the sari which was always attractive. I am sure this was much appreciated by our passengers. Not only was the cabin crew's standard of English exceptional but also they had a very nice manner with them.

Air India Crew arriving 'home' in Mumbai

I know that they really liked to fly on the Flyglobespan aircraft because they got on so well with the cockpit crew. I was told it was far different in mainline. There they were treated like second class citizens by the 'Little Maharajahs' sat in the flightdeck. We just used to treat the cabin crew normally like we would in Europe with an airline that uses modern CRM. As far as we were concerned there was nothing special going on, but they saw it differently. I liked to find out where they came from and some information about their home life when we chatted.

Not infrequently we flew the 767 as two captains together. Especially when one of the captains was new to the Indian contract line operation. It did have its quirks. I recall I flew one trip with a very experienced training captain called Chris. We had similar backgrounds in our previous flying careers and much in common. It was no surprise that we got along well. We had a similar sense of humour and some of the stuff he came out with made laugh out loud while we were flying. For example we had met before while we had been working for different companies.

I had been a member of the interview panel in easyJet when Chris had applied for the position of Head of Training for that company. The funny thing was that even though he was the best of the five on the shortlist of applicants, he didn't get the job. He was naturally interested to learn more, so I told him. Not only was he the best candidate from my point of view (and that of others on the panel), but he was head and shoulders above the rest.

We were informed however by our boss that one of the *internal* candidates was going to be awarded the position. I was saddened to realise that our efforts to run a fair competition with a level playing field had been in vain. It was obvious to me that the decision had already been taken by the senior managers above me before we had even started.

Such is the stuff of office politics and I was at pains to reassure Chris that I had not known the selection process was a waste of time until the 'wash-up' at the end. I was not surprised when the newly appointed Head of Training did not turn out to be ideal for the role later. Chris on the other hand, had gone on to greater things with his own airline and had become the Head of Flight Operations there.

Now that is a serious position in the hierarchy of a British company, also, it is well respected when you see it on somebody's CV. Think about it. You are the designated postholder for the UK CAA in the Flight Operations department of a huge airline. It is an immense role.

I had heard a rumour though that Chris had left his post in a hurry. He had resigned quite suddenly. As this was an ideal opportunity for me to find out more and I had already been candid about his selection process at easyJet, I asked him straight out why he left. I could not have predicted his reply.

"I decided to experiment with how many times I could say 'NO' to the managing director," he replied with a wry smile, "I found out it was ONLY ONE!"

I recall our laughter even now. We clearly held the same views when it came to airline office politics.

Air India - Birthday celebrations in the Forward Galley

On that particular flight after we were established in the cruise and heading west into the sunset towards London, the Senior Cabin Crew Member (SCCM) came in to the flightdeck to get our choices from the crew food menu. As we had not flown together before and all of us had been very busy prior to departure, she sat down on the jumpseat to chat for a while.

Dressed in her sari uniform, she was a very beautiful advertisement not only for the airline, but also for her country. We had not flown together before, but as I mentioned previously, the Air India mainline crew were delighted to fly with us. They enjoyed the respect we showed them, which contrasted with the verbal abuse they sometimes received from local cockpit crews.

I struggled to pronounce her rather long and difficult name. She appreciated its challenge for our Western tongues.

"I have a nickname," she said, "Call me Jugs!"

We pilots looked at each other. Chris appeared to be in some pain, suppressing a fit of the giggles! He had tears in his eyes, and I pursed my lips and shook my head thinking, *now – where - do we go from here?*

I hesitated to answer her, not trusting myself to speak at first, so she grabbed the crew menu to return to the cabin.

"I'll come back a little later for you guys."

"Okay Jugs..." and she left.

Now, we were on our own again. Chris was very amused and so was I.

"No James, we CAN'T call her Jugs, For God's sake man!"

"Well - SHE'S the one who said, 'CALL ME JUGS!'"

His shoulders were shaking and he was now actually crying laughing. So I asked him what we should do. We decided that I, as nominated aircraft commander, had the responsibility to inform her what exactly 'Jugs' meant in slang English. When she returned to the flightdeck I asked her to take a seat and please not to take offence, but I had something serious to tell her.

Firstly, I apologised for not telling her earlier, because I really had been taken aback when she told me her nickname. And she had left the cockpit before I could address the situation. Thankfully she saw the funny side of it and did not seem overly embarrassed by her gaffe, which I was pleased about. Heaven knows what she told the other girls in the forward cabin though. Whenever I saw her after that we always smiled at each other. It remained our secret.

✈

Not all the FGS people had such a good rapport with the locals though. One guy who practically lived in the Leela Hotel had been appointed in charge of the operational support of the aircraft. We pilots wondered why the Indians tolerated this guy's behaviour. He was rude about the locals, not only behind their backs but to their faces also. We were frequently ashamed of the shocking way in which he referred to and treated them.

We always spoke up in the defence of our hosts but this guy continued to abuse them all. Not only that, he had the approval of our senior managers in Edinburgh to act as he saw fit in the down-route management of the operation.

None of us were surprised when he finally got his comeuppance. I attribute it to Karma. His job required that he sometimes travel on the jumpseat of the aircraft during our flights. This positioned him to audit and monitor the flying operations on behalf of Edinburgh. To be frank though, we hardly ever saw him on the line, as he was loathe to leave the confines of the Leela. He seemed to spend most of his days there in the bar telling jokes about the Indians.

One day however, he did come with us on a trip to Dubai. He had to have a passenger ticket for our flight to leave Mumbai and it was his declared aim to simply "grab as much duty-free as possible" before returning on the inbound service. On the way back to base, I noticed he had changed into the uniform of an aircraft engineer but did not think anymore of it. Like I said, he was a law unto himself. We let him get on with it provided he didn't interfere with our flying.

After the flight he vanished and was nowhere to be seen when we returned through Mumbai airport. Certainly he was not on the GenDec (Customs/Immigration General Declaration form) for the crew. What we didn't know was that he had dressed up in his uniform so he could breeze past the immigration/border police quickly without having to wait in line with other passengers for passport checks. You might imagine, having the patience to wait in a queue was not one of his strong points.

Well this was all okay until the airport carried out their headcount at the end of the day. It seemed they were one inbound passenger short. They recounted again. And again, and again. There was no doubt about it, one passenger from an inbound flight to Mumbai International Airport on that day had not presented himself to the immigration police. This left the police scratching their heads, until finally there was only one answer. One person who should have come through the passenger channels entered the country illegally as a crew member – you can guess who it was.

They identified him from the CCTV footage which took some extra time apparently, and now many of the police were working overtime to solve the issue. Some of them should have been off-duty hours ago. They were not best pleased when they arrived at the Leela Hotel in the early hours of the morning to look for their illegal entrant. Neither were they impressed by the response of,

*"YER CAN F*** ORF!"* when they called the room number asking him to come to the hotel reception. Shortly afterwards the police went to his room and banged on his door until he opened up. They then proceeded to drag him in his pyjamas and with all the contents of his room down to reception. Here they dumped his possessions in a heap and at 4 am local time, officially arrested him for immigration violations. All the while he was protesting and swearing loudly at all concerned. He was then put in the back of a police van, deported and we never saw him in India again!

Lest you think that the bureaucratic processes of India are deficient in some areas, when it comes to who enters and who exits the country they can count! We pilots were all quite impressed. We never realised that they went to the trouble of counting the exact numbers of people in and out. You would have thought when a nation has a population of more than a billion, then a few more either way would not matter. Apparently it does.

We did see him again, but that was in Heathrow, still working for Flyglobespan. There was a rumour that he was erasing QAR recordings to delete evidence of tail-scrapes on the B767 fleet... but of course, THAT was only a rumour. True to say he was not the best ambassador for the company.

✈

Back in Mumbai the operation was going well. We had to adapt our thought processes to be flexible which helped. Is that the right way to put it? I think so. For example, from my time working out in the Far East, you simply had to make sure you didn't take things personally. When interacting with westerners, it seemed that the most important thing to them was not to lose face. 'Loss of face' is of huge cultural significance and so rather than tell you what they know to be true, they tell you what they think you want to hear. I have seen this with ramp personnel all over the world. If you ask them a question they give you an answer that later turns out to be false but you never see them again.

One of the advantages I found with being *based* in Mumbai was that we saw the same faces day in, day out. I am always eager to learn names... (no, not Jugs. Don't start me laughing again). For example, I would see one of the gate agents and say,

"We are ready to board Mister Pradesh, WHERE ARE our passengers?" and he would smile and wobble his head from side to side,

"Oh Captain! The passengers are coming!"

"NO Mister Pradesh! They are NOT coming," I would reply, "you told me that YESTERDAY! Now go up to the gate and FIND THEM! THANK YOU!"

I was always scrupulously polite when talking with Indian staff. I was aware that if they saw me become angry then they would think less of me. "Captain loses face Lah!" This would mean they would become less helpful. Whereas if I was straight with them and called them by their name, asking them specifically to do something, usually it got done.

"WHERE is the fuel truck Mister Gupta?"

"Oh Captain, the fueller is coming," came the predictable answer.

"PLEASE Mister Gupta, go and CALL THE FUEL TRUCK AGAIN, he is not coming, is he?"

Off he would go and a little later the fuel truck magically appeared. Naturally there was always somebody to ask to help. The airport looked to be overstaffed in all areas. I guess working for the airlines is a really good job in Mumbai. There were times when you would still be surprised by things however.

As a matter of policy and to reduce ambient noise levels on the ramp, we used to utilise the Ground Power Units (GPUs) to provide electrics for the 767 while on the gate at the terminal. This would also save us fuel burn for our APU, but it did mean that we are at the mercy of the groundcrew in a way. It was not unusual for the whole plane to go dark suddenly as all the lights went out as the electrical Ground Service Bus failed.

We would rush then to protect the aircraft batteries, as that was all the power we had left. Sometimes it was because the connector plug had fallen out of the socket in the hull, or the GPU had gone tech. The funniest I had was when the Dispatcher came back up having checked the GPU to tell me it had run out of diesel! When I enquired further, he said they were only allowed a small amount of diesel each time they went to refuel the GPU in case somebody stole it! Well that's what he said and I had no reason to believe otherwise.

I had already witnessed other irregularities during my time in Mumbai. There was the time we came in from the UK late at night and enroute to the Leela, we could see that the airport approach road had all been dug up. They had started the project prior to our flight out the week before, so I was not surprised to see the work was still ongoing. What did surprise me though was the police were actively guarding the roadworks even at the late hour we arrived. It was clear they would be there all night. When I asked our driver why the police were there, he looked a bit sheepish,

"The people have been stealing the road!"

Crazy, but true. The local people had taken to stealing concrete from the piles at the roadside to help with building their houses! I shook my head. I had not heard that one before.

Then there were the technical idiosyncrasies of flying airliners in India. Like some of the strangely written instrument procedure Jeppesen charts.

Bear in mind these charts depict required tracks, altitudes and distances which airliners must follow. They are intended to be used for flying in cloud on instruments. So, when it says, **"at 1700' and after crossing coastline ...."** you can imagine our amusement and confusion! After all, the question is, 'Okay if we're in cloud, how would we SEE when we had crossed the coast?' In reality we just smiled and found a workaround solution for issues like this. A simple range ring on the Fix page in the FMC was enough. There were many others of course.

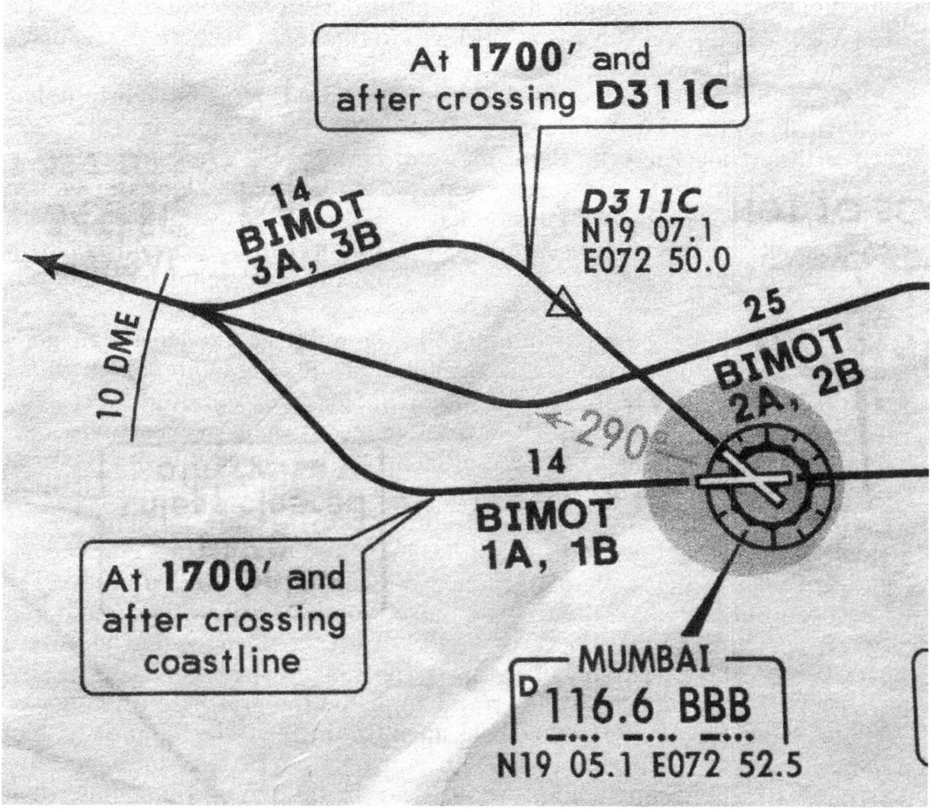

Jeppesen Standard INSTRUMENT Departure chart for Mumbai

✈

On another arrival into Mumbai, my company phone was ringing as soon as I had switched it on after shutting down the engines at the gate. The voice from crewing was stressed,

"James? What is the soonest you can come back on duty!? We need you and your FO to operate a flight out and back from Mumbai tonight."

I calculate 10 hours in the hotel and then give them a time. They agreed and said they would send me the flight details later. Meanwhile I was left trying to imagine what had gone wrong behind the scenes. I knew from experience that it was an unusual request. Apart from anything else, our contract said nothing about flying in-country. I concluded that this favour they were asking was to dig somebody out of a hole. We did not fly domestic sectors for the client airline – they were doing that with their short-haul fleet.

When we checked into the hotel, we went to our rooms and I advised the Gavin the first officer to get plenty of rest. He would need it; we would be up all night. I received an email from the local flight planners informing me that the flight we were to operate tonight was to Cochin in southern India. The IATA three letter code for the airport was COK. It was to be there and back, two hours each way in the 767.

To be honest, I had never heard of the place, but apparently it has historical significance as the centre of the world's spice trade for centuries. I was a little concerned the flight may not be legal because we were a UK flagged aircraft, so I needed reassurance from our B767 Fleet Manager in Edinburgh. Provided he said it was okay in writing, then I would be fine with it. So far all I had received was a panicky telephone call from a young chap in crewing and then the briefest of emails from the local agent handling our contract.

I knew we didn't have much time and I wanted to get the Fleet Manager's attention, so I titled the subject of my email; *'Now, The Flying Programme is Going to COK!'* – I thought, *that should get his attention.* And I was right, he replied within 10 minutes. He checked and confirmed we were good to go in his reply to my mail. That was sufficient for me. I met up with Gavin in reception in good time for the evening departure. Neither of us had been to Cochin before, in fact this was Gavin's first trip to the sub-continent.

We reported for duty at the Air India Ops centre with its crazy clock again, and got all the flight planning done. We checked Cochin carefully. The weather would be okay, the runway looked long enough, wide enough and strong enough – I cross-checked the PCN with the dispatcher. There were even some radio navigation aids at the airport – whoopee!

Then we met with the cabin crew, completed the pre-flight briefing and headed out to the jet. We pushed back ahead of the revised schedule they had given us. I noted that although Gavin and I were a little disoriented after arriving straight from London, some sleep in the wrong time zone, then flying again, we were working well as a crew. Lucky for me, he was an excellent FO and we had flown together before, so we made a good team. When we finally reached cruise altitude we were able to relax a little. Over a cup of tea, I asked him what he thought of India so far. He thought about it for a second,

"WELL... it's not quite what I EXPECTED!"

His response made me laugh out loud. That was a perfect description. Shortly thereafter the ATC controller asked us why we were routing down a particular airway. Not in the mood for any long discussion I brusquely replied,

"Air India 123. We are following the flightplan issued by the company

Operations Department. It was filed with ATC and approved. If there is another way you want us to go, then give us the reroute and we will fly that one. We just do as we are told!" He didn't ask any further questions after that, so we continued the way we were flying which I noted was the shortest possible route to COK.

The turnround in Cochin was uneventful. We got the passengers off and the new ones on again as soon as possible and trucked off back to Mumbai. I noted as we taxied in that it was getting light, we had been up all night, yuk! Thank heavens the Leela Hotel would be waiting for us – I couldn't wait to curl up in that huge comfy bed. Then later we could all meet up for an 'adult beverage' in the bar, before sitting down to a fabulous curry dinner. There were some plus sides to this detachment after all.

The contract lasted for another year or so after this. By then the client airline had managed to train enough pilots to fly their 777s. Our airline's contribution to the fly-pro was gradually wound down. By that time I had moved on and was flying in another part of the FGS network. I was on the South Atlantic Air Bridge operating to and from the Falkland Islands through Ascension in the middle of the Atlantic Ocean. But that is another story altogether. For the record, I would happily go back to either India or Ascension Island – they are both amazing locations, with great people.

✈

BACKSTORY: Mumbai was also a great place to get your advanced cosmetic dentistry. Dr Neetika Modi had a long list of pilots coming to visit her in Juhu, West Mumbai. Exceptionally skilled, she was dentist for many Bollywood stars. http://mumbaidentistry.com/ *Also, her father was one of the most senior officers in the Indian Air Force.*

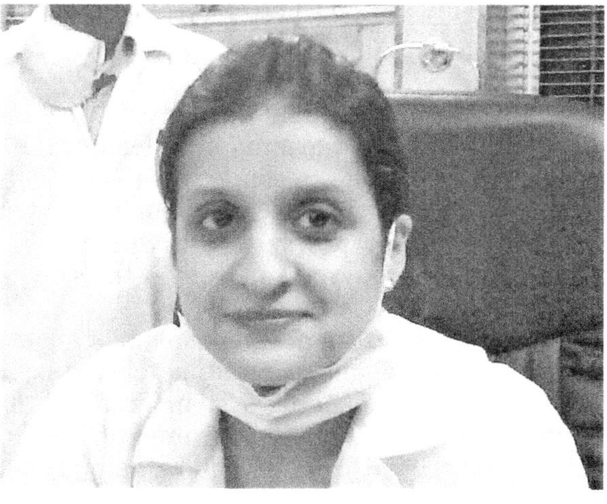

Dr Neetika Modi – expert dental surgeon to Air India pilots

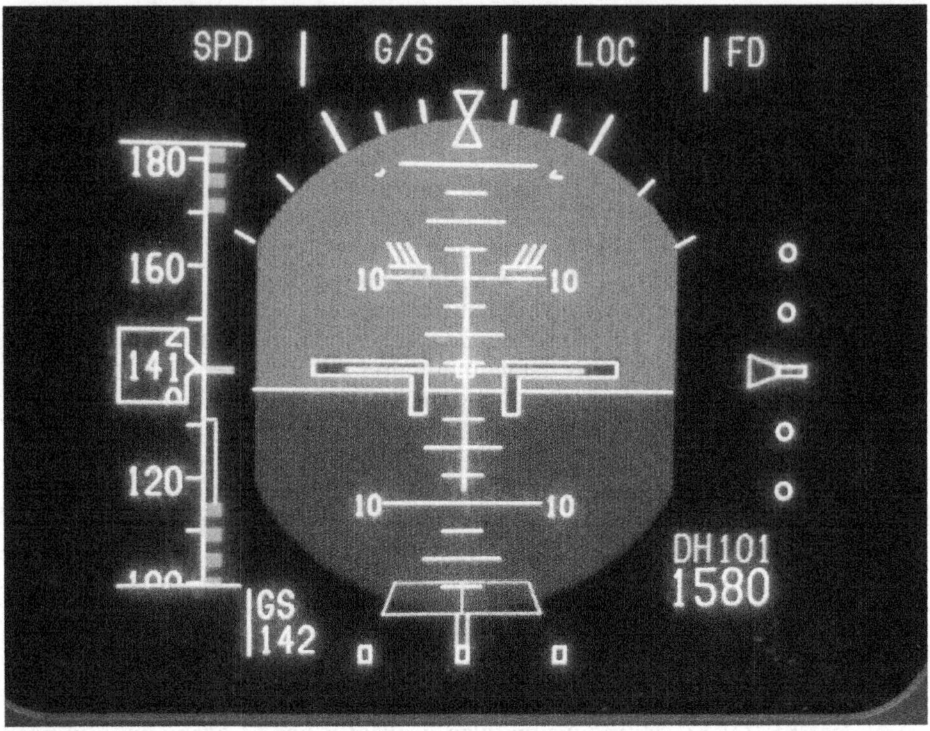

B757 – ILS display on EADI – on centreline, on glideslope

24 FLYING THE PERFECT ILS

It is often said that the perfect flight doesn't exist. Yet those of us in the commercial world are encouraged to aim for perfection at every level of our flying. Perhaps that answers the question 'Isn't airline flying boring?' I always wonder at this one, and after much thought have determined that this question is only ever asked by people who have never done it. So many variables make no two flying days alike, even on the same route in similar weather – and this is the key.

In the same way that professional comedians talk about looking for the perfect joke, most of us airline pilots are ultimately in search of the perfect flight. The ILS, or Instrument Landing System approach, has been around for 60 years or more and, together with instrument flying, still serves as the cornerstone of safe and regular commercial flights in all weathers. Precise radio beams from the ground are aligned with the optimum approach path for the runway. The aircraft receives them, displays deviations from the ideal on the instrument panel thus providing a guide for the pilot. Sounds easy?

Well it is not. It can be a challenge even for the most experienced aviator.

The Autoland system in all the larger commercial aircraft uses the same radio beams. In my view the perfect ILS is flown by the autopilot all the way to touchdown and monitored by two committed professional aviators who ensure that the system meets all the performance parameters as designed by the aircraft manufacturer, but Of course that's not the whole story is it? The secret of flying a good ILS approach starts miles away from the airport, even before you take off. Assuming you have checked that the ILS is serviceable and available and that you have the required approach plate for the airfield then you must consider the weather. Really close analysis of the forecast weather conditions at the destination airport at the expected time of arrival will reveal several important facts.

Firstly cloudbase – how close will it be to the minima? Is a Go-around likely or is airborne holding a possibility? More fuel will be needed to account for this. Time spent at the planning stage is never wasted. Ensure that you study the approach plate carefully, especially if flying single crew. Many years ago one of the pilots at the air taxi company where I worked returned from Gatwick having just completed an approach there in poor weather. It was a localiser only approach (no glideslope) in a light twin and he got in even though the BA flight ahead of him and the Air France behind both went around! From our subsequent discussion it turned out that he didn't know what minima to use for the approach when the glideslope was unserviceable – very scary!

The message here is know your minima and stick to your guns. Anything else could result in tragedy. As part of the weather analysis, the wind plays a big part in how to fly the ILS approach. Remember it decreases and backs in the latter stages of the approach, so you must take this into account when planning what heading will hold the localiser. The drift will be different reduce between the 'last platform altitude' (possibly 2,000-plus feet AGL) and short finals; therefore it will be more susceptible to drift on short finals as you slow down. Planning for this in advance will make the flying of the approach much more accurate, regardless of the aircraft type.

Groundspeed becomes a major consideration when deciding how much power will be needed on finals to maintain the three-degree glideslope. A 25-knot headwind with a single engined aircraft, could result in a final approach groundspeed of only 60 knots. Using the time honoured 'five times groundspeed' calculation this would require a rate of descent of only 300 feet per minute as opposed to 450 in still air. The same principle applies to piston twins, turboprops, executive jets, airliners and even the Airbus A380 – it's all a matter of applied physics. It doesn't take a genius physics boffin however to make a plan of the required vertical speed to scan during the approach in advance. Just a thinking pilot staying ahead of the aircraft.

Now we can think about technique. We are discussing manually flying the approach here, more of the automatics later. 'Selective radial scan' is what the best instrument fliers use to control the aeroplane while operating in IMC. This means using the principle of the Attitude Indicator as the master instrument.

Most of the pilot's focus is on the master with glances outwards (radially) to check the other performance and control instruments such as ASI, VSI, power setting, heading and of course the ILS needles themselves. It may not have escaped your notice that the ILS needles come last on the list, because if all the rest is correctly set up and monitored, then the aircraft flightpath should follow the ILS as if it were on rails. It is possible to fly a very accurate ILS by chasing the needles; but that's all you will be able to do. You will have nil spare mental capacity for managing the rest of the flight safely.

The SRS technique means that you will be able to think and fly at the same time – very important especially if you are flying the approach manually on your own. Minimising distractions is a vital aspect of the management side of things. For example, you should get into the habit of simplifying your R/T so that only the key words are transmitted – when handed over to Tower frequency, keep it short and precise. If the Tower clears you "G-XX Continue only. Expect late landing clearance", your response should simply be, "Continue, G-XX." Reading back just the proword is essential for two reasons, it keeps the frequency clear of surplus transmissions and ensures that you as the operator are not distracted from job one – flying the aircraft!

As part of staying ahead of the aircraft, most of us preset the next expected radio frequency in the standby window ready for a one switch change on handover, when transferring from one ATC agency to the next. There is a time when this is not a good idea and that is on final approach when you are talking to Tower – always have in mind that a late Go-around may result especially if you are not visual at Decision Height (Decision Altitude if flying a QNH approach). In the event of a Go-around the one frequency that you will need in a hurry is that of the Approach controller. There is no rush to chop to Ground control after landing when moving at only 20 knots or so, contrast this with the situation during an unplanned Go-around when you are working quite hard to operate the aircraft systems and you are climbing/ accelerating towards 200 knots at the same time!

Having said earlier that the perfect ILS is one flown by autopilot, being monitored and operated by the human element in the cockpit, you might think that all automatic approaches are the same. You would be wrong. The circumstances surrounding each approach make some more perfect than others. Like back in 1993 while flying as a captain on the Boeing 757. We were sub chartered to operate with a callsign of 'FunJet 123' for a Spanish carrier. Where do they get these names from? Our destination was Helsinki in Finland from the Canary Islands which was a flight of some five hours or more.

During preflight in Las Palmas, I asked Rob (the first officer) how much fuel he wanted to take, as he was flying the sector as the handling pilot. His response took me a little by surprise because he nominated a much larger fuel load than I had in mind. He had made a careful study of the Helsinki weather and reasoned that the early morning fog might hang around longer than forecast, so we should be prepared to make an Autoland there.

We had two diversion airfields on the flightplan. The first was Stockholm which was a long way away from destination and required a lot of fuel, but where the weather was excellent. The second was an airfield called Turku – quite close to Helsinki, had a similar weather forecast but it remained a viable option.

After five hours flying, we were in the descent and advised by ATC to slow down as 'some holding' would be necessary. The expected frontal clearance had not arrived and the fog was as bad as ever. We would have to make an automatic ILS approach. Our first holding pattern started at something ridiculous like 15,000 feet. Well at least the fuel burn will be less up here I thought, now very conscious of our fuel situation. The controller was very busy and there were lots of flights holding for Helsinki. His words.

"FunJet 123, take up the hold you are Number 21 for approach!" did not instil any confidence that we would be arriving in Helsinki any time soon. As Rob was flying the aircraft, I was busy with the electric slide-rule, calculating and recalculating the fuel situation. We had been holding for some 15 minutes and were about to go below minimum fuel to make Stockholm if we had to Go-around – it was decision time.

The air traffic controlling was very efficient and soon we would be able to make our approach. The weather was right on the minimums for an Autoland, (in Category 2 conditions – 100 feet Decision Height on the Radio Altimeter) with fog and low cloud. It had been snowing earlier, but the runway was reported clear. I decided to just check with Turku, now our only alternate to see if they had received any aircraft in from Helsinki.

"Off the ATC frequency Rob, I'll just check the Turku latest weather." The response from Turku was not good. The weather was right down to the limits now and they replied.

"No we have not received diverting aircraft from Helsinki, but some have diverted from here!" Ouch! This was exactly what I did not want to hear. Already we were committed to using them as an alternate airport. My stress level went up a gear.

Either way this is going to be very close – I thought to myself. As we left the hold with radar headings for Helsinki, the controller advised.

"You have 40 track miles to touchdown..." Phew, not even a shortened routing for us – we will have to delay putting flaps out and gear down to save fuel. Rob was well up to speed with the aircraft and was handling the automatics with assured competence. All the time, I was thinking to myself, if we don't get in from this approach it will immediately become a Mayday situation, as we had minimum fuel reserves and our only possible alternate airport had poor weather down to minimums...

The aircraft locked on to the ILS beam exactly and flew down the glideslope with total precision.

"One thousand feet..." nothing seen, complete clag outside. "Five hundred feet..." still nothing to see outside, I am nearly holding my breath now mentally willing the fog to clear.

"One hundred above..." STILL NOTHING!" Oh please... "Decide?" – fantastic, two reds, two whites – we're visual,

"LANDING!" I all but shout the executive command as I take control for the touchdown on the frozen runway. The main gear touched down in just the right place and the nosewheels thumped all the runway centreline lights as we decelerated towards the turn-off under autobraking.

"Good girl, Oscar Julie!" I said as I breathed a great sigh of relief. As the aircraft slowed to taxi speed, the computer came up with the message on the screen 'Insufficient Fuel'. Yes – when I think back now, some 28 years later, that really was the perfect ILS.

BACKSTORY: This story originally appeared in Flyer Magazine in 2007. Thanks to Ian Seager (Seager Publishing) for permission to republish it here. Check out https://www.flyer.co.uk/ *described as the UK's Favourite General Aviation Magazine.*

The FMC "Insufficient Fuel" message indicated that we were right down to our minimums. If we'd had to Go-around from that approach, we would be urgently looking to land straight in at our alternate airport. On this day it worked out, but we usually would not cut it that close. I was forever grateful to Rob for nominating a good amount of extra fuel, on that day. I learned about fuel planning from that for sure.

A very similar view of the snow-covered runway in Helsinki

25 THE DAY THE AIRLINES CAUGHT A COLD

When your planes are grounded the fixed costs remain

Without a doubt, the COVID-19 Pandemic is the worst global public health crisis to occur in a generation. Sadly we in the airline industry have inadvertently played a large part in the spreading of a virus which if not for easy access to intercontinental flights might have remained in China. Not only that but it might have been contained within Hubei Province, possibly within the boundaries of the city of Wuhan. It was not to be. Large multinational companies and governments have substantially more to consider than simply the rate of infection and how many people may die from an infectious disease.

Therefore international flights departed from China on schedule, some of them carrying infectious travellers. These went on to infect others either enroute or at their airport/country of arrival. Perhaps the cruellest aspect of the virus is that it can be carried unwittingly by an infectious sufferer for five to 14 days BEFORE they experience symptoms themselves. *Carriers can be asymptomatic and infectious for a week or more.*

In our industry we have always placed great store in the question, 'Are you fit to fly?' In fact a Medical 'Fitness to Fly' Certificate is the document issued by a physician that crew use to determine that the passenger can safely travel. Fitness to Fly covers a multitude of potential illnesses and infirmities. In every case, it relies upon the individual realising that they are or were sick.

Now they may have boarded the flight/s from China without any feelings of being unwell and flown to their ultimate destination far away across the globe. Then they travelled around meeting people, pressing the flesh, (like we don't do now) for days before finally succumbing to illness and taking to their beds. Meanwhile, the airliner used for the journey, may have contained infectious elements of the virus (perhaps on poorly cleaned tray tables for example) causing more passengers to pick up the disease.

Recently it brought it all home to me (literally) when I returned from New Zealand all the way to London in under 24 hours. Naturally all were aware of the coronavirus prior to departure. By that time I think the savvy majority were being scrupulously clean regarding hand-hygiene and potential contact zones. As I walked through Dubai terminal two weeks after the last time I was there, the numbers of people wearing masks and in some cases gloves, had trebled.

I thought about the length of time that it once took for travellers (emigrants for the most part) to make that journey to the New World of Australia and New Zealand. It was three months! Back in the 1788 the first large fleet of 11 ships landed in Botany Bay, Australia fully three months after leaving England. If you had been ill with an infectious disease during that voyage, then, by the time the ship docked you would have been dead or recovered. And in either case, no longer infectious.

Wind the clock forward 200+ years and ordinary citizens have the facility to cross the globe in a single day. My flight from Christchurch to Dubai was 17 hours. Then after a couple of hours on the ground another eight-hour flight brought me to London. And of course, that was the way the virus spread so quickly into Europe. With it of course came the economic consequences. Now the airlines have so many unfilled aircraft they cannot fly. But ATC Slots must be used... use it or lose it!

So consequently the airlines are flying empty aircraft to fulfil the legal requirement of 80 percent slot use. I know it is hard to believe, but there is pressure on the airlines from ACI World, the body that represents the airports. They are insisting that the companies abide by the legal obligation to fly their slots even though they have no passengers – how crazy is that? I flew some sectors last week for a scheduled airline all the flights had been booked full. We have 190 seats on the aircraft, but on some of the flights we had only 30 passengers. Now the airline is cancelling flights some two or three weeks ahead. Of course, that is logical. If there are no customers, why fly?

Every day brings a new story. Now the U.S. president has barred passengers entering the United States from Schengen Europe, claiming EU nations failed to contain the spread and noting virus cases there had outnumbered those in China. He has a point of course – if you want to stop people from infecting each other, then stop them meeting. Airlines are under severe pressure and some will no doubt perish.

Financially the airline business operates on ridiculously low profit margins anyway. It is a constant surprise that any airlines continue to post profits in a normal operating environment, let alone one where there are such challenges. That being said, the industry has survived many difficulties/challenges in the past decades. In no particular order, we have had the fuel crisis in the early 1970s, two Gulf Wars, 9/11 and then Afghanistan, the 2008 global financial meltdown, the Arab Spring and the collapse of Libya. Through it all, we have kept going and the industry has grown year-on-year.

As a trainee pilot, you could be forgiven for thinking that you have chosen the wrong career. Everywhere you look there will be bad news stories about what has gone downhill with the airlines – don't forget that nothing travels faster. (I'm talking about the bad news not the aeroplanes). However, the airline network is vast and the number of flights taking off and landing every day is huge. This current hiatus is exactly that. This too will pass... Mankind will not go back to taking three months to travel between England and Australia. There will be new aircraft designs, new routes, new customers in the future.

In the past 30+ years I have seen the airline industry go from feast to famine regularly in approximately seven-year cycles. First there are not enough pilots, then the companies train too many! It has always been this way. However in recent years I have seen a consistent shortage due to increasing barriers to succeeding as a commercial pilot. In the United States they have raised the bar; now a minimum 1500 flying hours is required. Due to the peace dividend, end of the Cold War, there was a dramatic reduction in the number of military pilots leaving the services and entering commercial aviation. Finally, it seems that the profession of 'Airline Pilot' is less desirable than it once might have been. There is no disputing that, with a whole raft of companies joining the race to the bottom in terms and conditions. Now there are many other professions that seem more attractive than 'Airframe Driver'.

We still need new blood. New enthusiasts to take up the challenge of keeping the standards high and the operation safe. Fresh eyes and ears to face the daily goal of Threat and Error Management. Some of us are on the last lap of our careers now, we are in the departure lounge. We know that the legacy of our generation is safe, certainly in Europe. "Safety First" was not just a strapline to us, not just a cynical marketing slogan. Sincerely it meant the difference between life and death for us all and I believe that even in these modern challenging times of economic hurdles, the world still needs to fly.

My message is one of optimism and confidence for the future and this despite the current state of viral uncertainty. If you are in the early stages of flight training, simply hold the dream. Focus on what is coming up next, don't worry about the advanced part of your training, it will come in good time. Keep working on progressing your Situational Awareness in addition to all the checklists and Standard Procedures. At the same time stay wide angle to keep hold of the big picture so you can see where all your effort is going.

Finally try not to take anything personally. Bad stuff happens from time to time, but don't let it cause you to deviate from your chosen course. If you want something badly enough, you will get there. Hold the dream!

✈

BACKSTORY: I wrote this piece on March 12, 2020 in Athens, two days after traveling across Europe from London. I could see the airline system was slowing down as all governments imposed travel restrictions. Little did any of us realise then just how devastating this industry recession would be.

Still, at the end of the global crisis, they will need pilots to fly planes, although in reduced numbers. I also suspect that the US and EU have abandoned any thoughts of raising the mandatory retirement age from 65 to 67.

Another less-visible cost of the crisis is the attention that now must be given to technical problems caused by prolonged aircraft storage such as insects in Pitot Tubes, birds nesting in landing gear bays. There have been many instances of aircraft returning to service with faults that have arisen as a direct result of long-term storage. Yet another challenge for our technical departments to face.

✈

Not chemtrails – Contrails, frozen water droplets and harmless

26 A GENTLEMAN'S FLAP SETTING

We can all agree that some days when we go to work, we perform better than others. Whatever the occupation, be it artistic, manual, skilled, unskilled, professional, white-collar, blue-collar – some days it all just clicks into place. On those occasions we are fulfilling all the requirements for our job and achieving all the required performance targets. I know when this happens as an airline pilot because I get this feeling of being 'connected'. Not just to the aeroplane (although that is a good place to start, strap into the seat and adjust it), but to the dynamics of the actual flying operation itself.

It is not easy to describe but there is a kind of free-flowing energy through both body and brain while you are part of all that happens. You are mentally way ahead of the aircraft and at the same time you experience intense situational awareness.

Recency plays a big part in this though. If you are returning to the cockpit after two weeks leave for example, then certainly you will be rusty and it will show. Your interaction with your partner in the flightdeck is clunky and slightly disjointed. The interactive checklists (Challenge and Response) are slow and sometimes hesitant with occasionally incorrect responses. Therefore we will usually prewarn our colleague when we report for duty after a period of time away from the job. This comes naturally as part of Threat and Error Management (TEM). We brief these threats separately before we get into the full departure and emergency briefing. The reason for this is that we nominate the threats first and mitigate them individually with our operational briefing. Forewarned is forearmed as they say.

On the subject of briefings, I have witnessed a tendency among the next generation of aviators to plan for the ***'worst case scenario'***. I lose count of the times I hear this expression in a working week. Its overuse nearly renders it meaningless. I cannot pinpoint the origin of this strange behaviour, but it has clearly become an infectious fashion. Maybe the first officers think that they will sound 'super-good' (another modern expression), if they manage to say these words in every briefing. Unfortunately it just sounds amateurish to more experienced airframe drivers. I interrupt their briefing (at a suitable time) to point out that we should really plan for **realistic** scenario's, having of course ***considered*** both worst and best cases.

Here is a case in point. When the TAF (Terminal Airfield Forecast) indicates bad weather, it is usually accompanied by a percentage of its likelihood. This is termed probability. So, a PROB 40 CB in the report means a 40 percent probability that there will be a Cumulonimbus in the vicinity of the airfield during the stated period. Clearly if the CB was located directly overhead the field or worse, on final approach to the runway, then this would be a worst case.

However, think for a moment. We are only going to be flying into the airport for a five-minute period during the whole of the TAF validity. Therefore the odds are very much less than 40 percent for us to experience this bad phenomenon during ***our*** approach to land. Sure if you were sitting at the airport for the whole day, then it is quite likely that you would see Cumulonimbus build ups for around 40 percent of the time. That does not connote 'worst case scenario'. Keep it real and keep it in perspective.

The airlines recognise this. In most companies' Operations Manuals, it will state that during preflight planning the flightdeck crew may disregard PROB 30 or less. This can be read and interpreted in different ways. I would say that provided the METARS* are supporting the accuracy of the forecast then it would certainly be safe to do so.

Why is the weather so important to us? Crucially it has a bearing on how much fuel we will order to be delivered to the aircraft. As we know to carry additional fuel will cost us extra fuel in increased fuel-burn. With modern airliners maybe it could be an extra 4 percent fuel consumption per tonne per hour. You may reason if you are flying all short sectors that this extra fuel burn is insignificant – but multiply by a fleet of 500 aircraft then you will see the cost involved.

Finally you should consider other factors regarding the destination airport, weather forecast and fuel requirements. One good example is Tirana in Albania. I have been there on three occasions in the past two years with the B737. On each visit we carefully studied the weather forecast in advance and found little of note. There were some isolated showers in the forecast during the period covering our scheduled arrival time.

It all looked innocuous to be frank. On the first two flights, the actual weather was considerably worse than the TAF. Seriously, it was pants! There were many cumulonimbus build ups in the local area and several thunderstorms with heavy rain and even hail. Our weather radar was working overtime while we picked our way through the obstacles, descending to final approach which was thankfully relatively clear.

The first officer and I discussed the inaccuracy of the forecast at length after we had landed from the first flight. On the next flight to Tirana only a couple of weeks later, I made a point of studying the weather very carefully. Again nothing indicated we should take extra fuel.

During the descent and approach, history repeated itself and we experienced foul weather. Lots of large build-ups, heavy rain, moderate icing in the cloud and plenty of turbulence. We concluded that the Met Forecasters in Tirana airport were either not very competent or experienced. The final time we went there I related the tale to the FO and we took plenty of fuel so we could hold for at least half an hour if we needed to wait for a heavy shower to pass through. Guess what? Same again. We were glad of the fuel.

I honestly don't think the Met men in Tirana deliberately tweak the forecast. I think they just get it wrong more often than their colleagues in other airports. But first prize for rewriting the TAFs, must go to Mombasa Airport in Kenya. Well at least they used to do this many years ago when we went there with the B757 full of charter holiday passengers. The weather forecasts invariably indicated a fine blue-sky day, but every time it seemed to be a real problem getting an accurate description of the conditions. It transpired that at the time the airport management had taken the view that if they told us the weather was bad in advance, it was more likely we would not come! True story. The guy on the Tower frequency would be calling out visibility in excess of 10 kilometres on the radio, when in fact we had trouble seeing the runway from a couple of miles on finals!

Like I said, there are times when you are flying the jet and you are completely on top of your game. Everything seems to fall into place nicely. Your descent and arrival inbound to the destination airport is going well. Ideally, we will carry out a Low Drag Approach, combined with a Continuous Descent Arrival. We aim to slow the aircraft down with minimum use of speedbrakes and only deploying the flap at the correct time as the speed reduces. When it is going well for you, it is a joy to fly. The first stage of flap in the Boeing 737-800 is Flap 1 and at this point you will ideally just be approaching the glideslope for the final approach with speed reducing below 200 knots.

It is usual thereafter to call for "Flap Five" and this will now provide some drag which prevents the jet accelerating as it intercepts the final three-degree slope. Sometimes however it is obvious that because of the weight of the aircraft and the amount of headwind at 10 miles out, if we took Flap 5 now, the autothrottle would add power to maintain the profile and speed. THAT is the point where you might call for "Flap Two" in anticipation of all the factors involved.

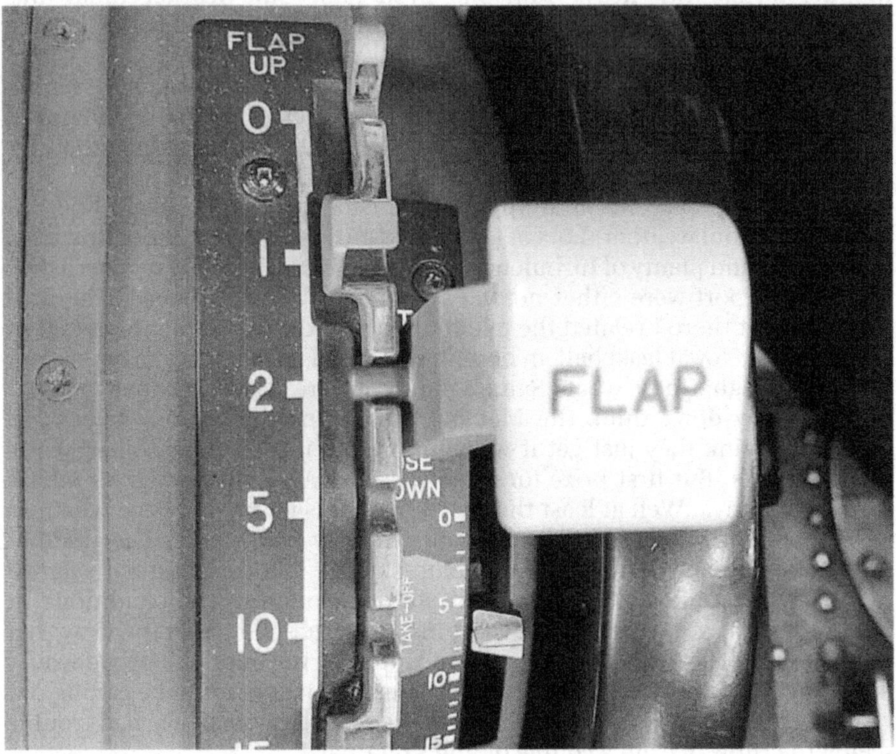

Ahah! A Gentleman's Flap Setting…!" may be said in a humorous tone by Pilot Monitoring as they move the flap lever for you. This expression signifies that they appreciate you have it all under control and are well connected to both the jet and the dynamic situation. It is a compliment really from your colleague. Although of course you should not let your guard drop.

While flying - Trust no-one, especially not yourself!

*METARS – Met Actual weather Reports

BACKSTORY: Maybe the Pilot Monitoring response call of "Ahah! A Gentleman's Flap Setting!" in response to PF's call for "Flap Two", should be made an SOP for B737 operators!

27 THE DEMO FLIGHT

VIP B737 – Ready for flight test

My main recollection of that extraordinary day demonstrating and selling a private jet was its logistical challenge. I was representing a private jet company that wished to bring another jet into its fleet. This time it was a Boeing 737, so not just any other bizjet, but a corporate airliner with 40 cabin seats. The aircraft had recently been refurbished with a new interior, was all certified and ready for operation. I was told that there was a prospective buyer who was interested. However he needed to be persuaded it was a good idea and could I help?

Naturally being a positive sort I was delighted to help. I approached the potential customer to outline the asset management services that my company could offer if he bought the plane. Our meeting was on a snowy day in the North of the UK and I had to drive over the Pennines to get there. From our first meeting, I found him a likeable individual with whom I was convinced we could do business. We discussed the potential for charter work with the jet and I suggested a nicely fitting market niche his company could exploit. In principle I could tell he liked the idea, but there was just the one sticking point.

"But James we haven't seen it fly yet. Does it fly? When was the last time it flew? I've bought a lot of expensive cars in my time and I NEVER bought one without a test-drive!"

I could appreciate his point of view especially as I was asking him to spend millions just on promises...

I admit to being caught on the hop here, I did not know when it last flew as it had been in the hangar for a very long time. So I countered with.

"If we fly it up here and give you a demo flight, will you buy it?" He didn't commit himself. Successful in business for a long time, he was not going to spend millions of pounds just on the nod. Although as we parted he indicated that if he had his 'test-drive', he would likely buy the aircraft and let my company operate it.

Now it just remained for me to get all the building blocks in place for a demonstration flight. I spoke to the managers of the aviation engineering company that had rebuilt the jet. They had already spent a lot of money putting it all together so they were not keen on the idea of giving it to me to me to fly for a few hours up North to demonstrate it.

"How do we know he will buy it James? If it's a wasted sales mission it will have cost us thousands with nothing to show for it."

I acknowledged this and finally we came up with a plan. They had zero budget for demo-flights, but I knew my company would foot the bill for the operating/landing fees and I was confident I could find cabin crew and pilots. This just left the expensive and daunting issue of insurance. They said that when they had requested a quote for flight cover, the premium required by the insurers was £20,000 for even a short period. That's what the airlines pay to cover operating assets worth tens of millions for a year, but seemed a bit steep for a couple of days. This had the potential to be a real "stopper".

I approached the insurance company that provided cover for all our company's private jets. Understandably, they were not keen either, but when I explained this was a demonstration flight for a jet that we would probably manage and operate, they saw it differently. My promise to place the policy with them in the event we became the operator sealed the deal.

This would also benefit the new owner because our company had established highly competitive terms on behalf of the other jet owners. They gave me a delightfully low price for two days insurance. Now we're in business, I thought to myself as I picked up the phone to call in a few favours. You see my company didn't have any Boeing 737 pilots and I had nil funds to pay for any.

A few calls later and that roadblock was dodged. I found a current captain and a first officer. They agreed to fly for the price of a few beers together and dinner. You know who your real friends are at times like these. I was still 737 qualified, but decided it was better for me to manage the project rather than operate the jet.

I would be there to look after and support the crew in every way possible so they could do their job. Bear in mind, my old chums were Boeing airline pilots and were used to having the back up of a large organisation. Filing flightplans, bringing pushback tractors, rustling up fuel trucks and passenger steps, organising onboard catering etc. I would supervise all these details. The VIP cabin crew were different. They were highly experienced at being resourceful and could be relied upon to sort out everything for a VIP jet down route in somewhere weird. I felt Bournemouth wasn't all THAT WEIRD - they would find their way.

The engineering company and potential buyer scheduled the date. By this time, I was having to communicate through his management team. They were quite protective, and I worried they would persuade him not to buy the jet. Oh well, too late for that I reasoned, we'll just keep going until something stops us. But it appeared nothing *would* stop us. It was all systems go!

Just to be on the safe side, I travelled to Bournemouth a day early and I'm glad I did. When I went to see the plane in the morning before the demo flight, I found that its last flight had been almost a year ago! The engineers were confident the machine would operate perfectly well in all respects for a sales demo-flight. While I applauded their faith and confidence in their own work and the integrity of the design, I expressed my concerns.

"How much more comfortable would we all be if we could carry out a test-flight today? Just to be completely certain that it all works as the manufacturer intended?"

They hummed and hawed for a while, then finally agreed. I had got my test-flight, but who would be doing it? My last 737 flight had been a while back, but I was still rated on the type, so I could hardly say No, could I? Especially when this had all been my big idea. So at very short notice I got another experienced B737 FO to join me in the cockpit. As I sat in the flightdeck it all came flooding back to me and I started to feel at home again. Back then, my current type, the 757, was very similar in a lot of respects.

I realised just how lucky we had been when the landing gear would not retract properly after take-off. We flew a big wide circuit, running the checklists and put it back on the ground again. We taxied in and there was a small reception committee of engineers ready to provide the fix. Sitting in the hangar for a year meant that one of the main landing gear microswitches needed adjustment. Just to be sure, after adjustment of the microswitches they raised the aircraft up on jacks and completed landing gear retraction tests.

All good, but we decided that we should complete a full check flight. Luckily I had carried out several functional check flights on Boeing 737s previously while working for easyJet. However that was for a large airline, with enormous technical backup. This was very different. For one thing, the engineers were concerned about how much fuel we would use. So we were under time pressure from the outset!

To say it was a challenging flight would be an understatement My colleague and I worked like one-armed paper hangers for 90 minutes airborne to get through the re-delivery check-flight schedule. Seriously we were full-on! At the same time, it WAS exhilarating. Here we were, pushing the boundaries to get the job done in the minimum amount of time in unfamiliar airspace, talking with 'London Mil.' Testing, checking, recording all the main systems and redundant systems. At the end we had a 20-year-old airliner that flew just as well as a new one. Amazing! Thank you, Mister Boeing. We had a couple of minor snags, but nothing we would not have accepted for 'commercial public transport operations' which was our benchmark.

In the evening the VIP cabin crew arrived with my demo flight pilots, Chris and Allan, good friends from a previous company. It was great to be together again, and we joined our three flight attendants for a very pleasant dinner. The flight attendants were more than a little sceptical when we laughed about the two flights undertaken earlier in the day. They were accustomed to flying in brand new private jets in mint condition. We were asking them to take a leap of faith here with a 20-year-old 'bus' that had never been designed for VIP operations. Give them their due, they got on with the task and came up with workaround solutions.

Still, I was left under no illusion that what I was asking them to do was really 'beneath them' - they were doing me a BIG FAVOUR! I reminded myself that these ladies were used to being treated like Sky Princesses. Five-star hotels were their norm, they travelled the globe, worked with billionaires and were a bit snobby about the three-star hotel I had booked in Bournemouth! I apologised profusely of course claiming it was all I could get at short notice.

I could tell by the way they looked pityingly at me that they considered me not in their league. Privately I had my own thoughts, but in the interests of good working relations I kept them to myself. By day's end my jaws would ache from maintaining a fixed smile on my face.

Whatever, all my planning was coming to fruition and I nervously prayed for a successful mission as I went to sleep that night. Early next morning it was so cold you could see your breath in the air as we loaded up my car in the darkness outside the hotel. I had the cabin crew in my car and Chris and Allan followed us in theirs.

Of course, I had started the car in advance with the heater on so the ladies wouldn't be too cold. They had been frosty enough the night before over dinner when we had laughed and joked about the **"old crate"** we would be flying today.

I brought the car to a stop in the predawn darkness next to a fence by the hangars. We could see the tail of the VIP 737 sticking out of the far hangar. The APU* was running which was a good sign. As I moved to switch the ignition off, the senior cabin crew stopped me.

"Just leave it running with the heater on and give me a missed call on my mobile when the plane is ready for us." I nodded my agreement, sucked my teeth and smiled.

"Yes, of course. We don't want you all standing around getting cold, do we?"

I fumbled my way through the combination lock on the gate muttering *"Lord save us from VIP Totty!"* as my two pilots joined me for the walk to the hangar. It really was freezing cold.

Inside we could see the engineers had been busy polishing and cleaning. They were still at it. While the flightcrew installed themselves and got cracking in the cockpit setting the machine up for flight, I went to the office to print off the weather and Notams. In addition I had coordinated with our Operations to file the flightplans. I called the VIP cabin crew to come too and I could see them leaving the carpark as I entered the engineering office.

Getting the email system to work took some time and I felt frustrated waiting for the flight paperwork to download and print off. But this was nothing to what I would feel in just a few minutes as I walked back to the hangar. With all the papers in a folder I headed out of the door. As I turned left on the concrete outside, I was suddenly struck by the absence of any APU noise. The tail was still sticking out of the hangar doors, but the APU was silent. This was not a good sign. Worse was the cloud of smoke drifting out of the entrance to the building! I started running.

As I entered the large open space, there under the arc lights stood the crew, outside the jet. Surely this can't be happening to me? I felt sick. Immediately I made the mistake of assuming that it was THEIR fault. The flightdeck DV windows were open and smoke drifted out of them – *Aargh!* I said to Chris the captain.

"What have you done to it!?"

He reassured me that everything was normal until the APU Generator Control box at the rear of the cockpit started smoking. They shut it all down and evacuated into the hangar. Fortunately fire extinguishers had not been required.

The engineers established very quickly that the source of the smoke had been a one-year-thick layer of dust that had overheated on top of the box. I looked at my watch - we could still do this.

In quick succession, the engineers replaced the box, the pilots jumped straight into their operating seats, the cabin crew quickly secured the unfamiliar cabin and galleys and we were taxiing out for take-off. Talk about CRM challenges and rushed departure! The poor guys didn't even have time to take their jackets off. We arrived at our destination airport only a few minutes late and ready to perform the demo flight.

The poor guys didn't even have time to take their jackets off!

The prospective new owner and his management team came onboard, and we showed them around the airliner. It did look beautiful inside. The 40-cream leather business class seats were set around Hi-Lo tables. All was perfectly polished and smelling very new. The cabin crew had done a good job, not just with the cleaning, but with the presentation of the VIP catering.

That said, I was still conscious that the 20+ year-old airliner had lived a very full life with the airlines. More than that I learned that the buyer had previously owned a brand new BBJ (Boeing Business Jet) and wanted to see if this machine a stock B737-300 would be as quiet in the cabin during flight. One of his aides told me he had downloaded a decibel meter App onto his iPhone before the flight and would be using it to check. I know I should have had more confidence, but the expression 'Putting Lipstick on a Pig' stuck in my mind.

All smiles now – mission accomplished

I had no idea whether this airliner would be as quiet in the cabin as his BBJ. But I was deeply disturbed to think the whole deal may be in jeopardy due to the perception of an owner with an iPhone. As quickly and as discreetly as possible I passed instructed the pilots to **"FLY SLOW!"** It didn't matter what happened once we were airborne, just so long as they kept the speed down and used correspondingly low engine power settings.

The experienced and aware pilots did exactly that. I don't think they exceeded 250 knots indicated airspeed throughout the flight. In the cabin, you could hear people even if they whispered! Which was just as well, because one of my crew overheard two of the new owner's managers in the rear galley discussing how they planned to allow us to operate the aircraft for them only.

"Until we can set up our own AOC operation, that'll only take around six months!" I laughed out loud when I heard this later. Setting up a UK Commercial AOC is a mammoth undertaking with huge pre-requisites. Seriously they had no idea!

When it came to the sound recordings and the measurement of the cabin ambient noise levels, the iPhone App indicated that our 'Twenty-year-old Bus' was actually quieter than his new BBJ. Something of a surprise to everyone!

We landed back in the rain in Northern England and the new owner and representative from the engineering company concluded the deal with a handshake. I breathed a major sigh of relief and thanked the crew profusely for their flexibility and professionalism. They were outstanding and we could not have got this over the line without them. After we said goodbye to our new friends, now customers, we closed up and prepared the aircraft for flight. It had been gratifying to manage the whole project through to its successful conclusion but better still was the satisfaction to fly the aeroplane myself back to base. I felt like a lottery winner!

✈

*APU – Auxiliary Power Unit in the tail of an airliner. Provides electrics and hydraulics on the ground when the engines are not running.

BACKSTORY: What an incredible team effort to get this one over the line. My thanks go to Shaun, Doug and Trevor at European Aviation, Bournemouth for their engineering excellence. Thanks also to Carl and Paul at Willis Insurance Brokers for their kind assistance. Plus of course the crew, especially my two buddies Chris and Allan who did me a big favour.

28 CONFIRMATION BIAS

The term 'Confirmation Bias' is a seemingly innocent combination of two ordinary words of the English language. Yet these words put together in the environment of professional aviation have a potentially lethal outcome. Numerous analyses of aviation accidents have identified key Crew Resource Management (CRM) failures that are common factors. One of these is Confirmation Bias. Always, while operating commercial airliners, we are aware of the presence of this highly dangerous phenomenon in our day to day working life. But why is it so dangerous? To really understand the threat you must consider that the safety and integrity of the operation relies heavily upon the operating crew maintaining good Situational Awareness.

This includes verifying all information which is received from outside sources. Indeed as individuals, we aviators can become a bit difficult to live with at times in our home life. We never take anything at face value. We are constantly cross-checking the accuracy of information received. We hate relying on any system which even hints at 'single-point failure'. We always have a Plan B. In short, we are often mistaken for being obsessive and compulsive. I know that my own family hate it when I ask questions like,

"What will you do if your first option is not available?" or "What will happen if your plan fails?"

Here is an example. Suppose we are out on the water sailing. When you are about to bring a boat into a dock which is fitted with fixed mooring lines, you know that you will not need to drop your anchor. Nevertheless some of us ensure we get the anchor ready for quick deployment anyway just in case the engine quits and fails to restart. If this happens, the first thing you need to do is drop anchor to give yourself time to fix the problem instead of being at the mercy of the wind and current. Call it contingency planning or healthy distrust of the mechanicals, but pilots always have a Plan B.

Which brings us back to Confirmation Bias and why we are naturally suspicious of information which arrives in the flightdeck from any outside source. History has taught us that we should be circumspect and cautious in order that we can protect ourselves from being misled to the wrong conclusion. One way to describe it is in the words of an old aviation joke which ends with the punchline, "Don't confuse me with the facts, my mind is MADE UP!"

Even in the earliest phase of learning to navigate aircraft with reference to ground features, we are warned that when uncertain of our position 'Do not try to make the ground fit the map'. It is essential to work with verifiable facts in a logical manner to prevent jumping to conclusions. Working as a crew we use many techniques to ensure that our mental model of the world is an accurate representation of reality. We try to avoid closed questions between us when flying the aircraft. We strive to create an environment which is conducive to any of the crew advocating their own suggestions in relation to our decision-making process.

My perfect example of Confirmation Bias occurred earlier in my career. I was working as a Line Training Captain for easyJet just after it had started flights. They say that hindsight is 20/20 and looking back, it is easy now to see how it happened. Although I cringe with embarrassment whenever I think of it, I am happy to share my mistakes so that all can learn from them. Bear in mind that the airline only had six aircraft in 1998. There were new routes being developed all the time and most days at work seemed to be ‘organised chaos’. I reported for duty in the evening for a return flight to Athens from London Luton airport.

This was my fifth day of duty and all my previous days had been long ones. It promised to be a long night with an extended turnround planned once we arrived in Athens. In the late ‘90s, the old Athens airport was still in operating. There were multiple challenges with all the noise limitations and performance restrictions. For example there was a ‘night-jet ban’ which prevented us from taking-off before six o’clock in the morning. Hence the ultralong turnround.

When I arrived in the crewroom the Ops controller called to inform me that we would have ‘performance restrictions’ taking off from Athens and if all the passengers and baggage were loaded as planned we would be unable to load enough fuel to get back to Luton as a direct flight. This surprised me because I thought that ‘tech-stopping’ for fuel was a thing of the past. However, this was in the early days of the airline and it seemed like everybody was learning, new things happened every day. There was little time for decision making and I was the captain.

“So Skipper what would you like to do?” I reviewed the options and looked at the performance figures carefully. Sure enough, we were about one metric tonne short of having enough fuel to get back home in a one-er. I also consulted with the trainee first officer and the safety pilot.

They had no better suggestions and we all accepted that our Operations people had looked at it. It seemed plausible in many ways. Athens, the destination airport was renowned for being tricky. High temperatures reduced engine thrust and required a longer take-off roll. We knew the dangers of engine failure on take-off in Athens given the published requirement for an emergency turn to avoid overflying the Acropolis.

The route from London to the Greek capital had only just started and although I had previous experience going there, that had been in the Boeing 757 which had better performance margins than the B737-300. From every angle, there seemed no way out. Using the maximum take-off power according to the performance tables we could not lift the weight. But more importantly to us it seemed, the guys in Ops had told us “EXPECT to Techstop.” Not only that, but they also stated we would have to change cockpit crew as we would be out of hours and unable to fly all three sectors after our extended ground time.

Time was pressing now and I had to focus on the training of the new pilot and ensuring he was adequately prepared and briefed to fly. We decided to stop in Geneva enroute to Luton to change crew and refuel. From memory I think this was Operations' preference because they knew they could get two pilots there easily from our London base. Unfortunately this flight was not going to make a profit. I shrugged my shoulders and we made our way out to the ramp. The cabin crew were unhappy about the situation, but what could we do? The Flight Time Limitations (FTL) scheme clearly stated that they could work an hour longer than the pilots, so they were 'in hours' to do the three sectors.

We agreed that we would call back to base after landing in to make sure that nothing had changed regarding the expected traffic load and when we had the actual Met. conditions. Bear in mind, when planning we are always working with the forecast weather which is sometimes better and sometimes worse than the actual. The departure and outbound flight were uneventful, albeit that the flight time was the best part of three and a half hours. After landing, we confirmed with the handling agent what the expected load would be and sure enough we could see the performance deficit. Ops assured the relief crew would be waiting in Geneva for the time of our arrival.

As we settled down to wait in the aircraft for the next three hours, I mulled it over in my mind. There was something which just didn't quite add up. We had all looked at the problem, reviewed the performance tables carefully and even taking every possible advantage into account, the aircraft was still short on performance. Oh, how I missed the beautiful 757 with its abundant power from those RB211 engines!

The hours ticked slowly by and we got more and more sleepy. We tried to snooze and rest before our next flight, but it was unsettling knowing that we would have to inform the passengers that sadly they would be seeing Geneva enroute to Luton. Not only that, but the weather forecast for GVA was heavy rain and low cloud with associated poor visibility - Yuk! Finally, the passenger boarding commenced with about 35 minutes to departure. We started to gear ourselves up in the cockpit and worked on the performance calculations.

Not an easy task. It was now the early hours of the morning and we were all suffering the effects of being at our circadian low. Again we ran the numbers and it was 1000 kilogrammes too heavy if we put on enough fuel for Luton.

"It's not relevant now anyway" I recall saying to my two less experienced colleagues, "we've planned for the techstop and crew change." Then it hit me like a train. "Hang on a minute! Where's the performance tables for a Bleeds-Off Take-off?"

Two minutes later, we located the big folder with the RTOW* performance for take-off without the bleed air pressurising the cabin. In this case of course, we would leave the APU running to provide the necessary air pressurisation. The engines, relieved of their load to provide cabin pressurisation would have significantly more power. And here it was in black and white, we would gain another 1400 kgs, so we could do it! We had very little time to make all the changes. We had refuelled a couple of hours ago and now we had passengers onboard. A call to the ATC Tower confirmed that we would need Fire Crew cover to refuel with pax onboard.

"You will need the big Fire Extinguisher and there will be an extra charge."

That didn't sound good. There would also be a further delay to get it here and cabin crew possibly going out of hours in that time? We would require a new flightplan. This would take time to file and send through for the handling agent to print off. And would it even arrive in time?

With so many variables we had to go ahead with the techstop plan. I cursed myself for not thinking of it earlier. We would fly to Geneva. I recounted all the above in the interview with the Chief Pilot a few days later when I was asked to explain why we flew home from Greece via Switzerland. To his credit he was very understanding about it.

The whole episode yielded valuable training feedback to the inexperienced controllers in Operations. The airline management was 'fire-fighting' on a daily basis and this represented just another minor crisis**.

For me as a supposedly experienced Training Pilot at that time it proved a salutary lesson in the dangers of Confirmation Bias.

✈

*RTOW Tables – Regulated Take-Off Weight Tables of performance figures. A much-outdated method of assessing airliner performance. In most airlines it has been replaced by electronic systems giving improved accuracy, safety and overall enhanced SA.

**Those early Athens flights challenged us, not least because of the length of the turnround. Our exhausted crews even became the subject to Prolonged Loss of Communication (with ATC) on the return flights and several airborne fighter interceptions.

BACKSTORY: This was not the worst fiasco we had during those early days at the Web's Favourite Airline. In no particular order, we almost had two hull losses, several needless diversions for technical issues that could have been resolved inflight, a few unnecessary diversions, some very odd crewing matters, strange decisions by operating captains on the line, plus equally inexplicable operations and management calls...

There was the 'Captain Rubber Jungle' incident, where on a short Amsterdam from Luton all the oxygen masks suddenly dropped down in the cabin. Nothing was heard from the cockpit, so the Senior Cabin Crew Member opened the flightdeck door, only to see BOTH pilots had donned their oxygen masks, but they hadn't TOLD ANYONE!

Flights carrying out high speed Rejected Take-Offs due to flightdeck window opening. Flights carrying out high speed RTOs due to movement of the captain's seat – not being locked in position for takeoff. You name it, we had it. Those of us who stayed with it came through smiling. Understandably, many pilots could not stand the constant changes, the lack of a stable roster and inconsistent messages they received from easyLand. In 1998/99, about 40 percent of the pilots left the company – and this was in a growth period when we needed to recruit more!

✈

29 CRM – NOT SO SMART

There are times when you just shake your head and ask yourself "What on earth were they thinking?" Like when you read the report of Smartwings Flight QS-1125 a B737-800 from the Greek island of Samos to Prague which occurred in August 2019. In summary, the B737-800 was 100nm northeast of Athens at cruising altitude, FL360 when the aircraft suffered an engine failure of the Number 1 engine. The crew flew a driftdown procedure to a cruising altitude (FL240) which they could maintain on one engine and to try to restart the failed engine. After two unsuccessful attempts, they continued the flight to land at Prague, two hours and 20 minutes later. The passengers remained unaware of the situation throughout. In fact, the crew did not inform any of the ATC units enroute that the aircraft was flying with one engine shutdown. In contravention of standard commercial aviation norms, they only declared the real situation when they were handed over to Prague ATC.

It is acknowledged that the actions of the crew and in particular the aircraft commander may have been negligent. Although this has still to be determined 15 months later. The issue which really stands out is that this happened in Europe. Airlines in Europe have a much-vaunted reputation for safe, well-regulated operations. The system works well to ensure that if passengers buy a ticket on an EASA registered aircraft, they do so with absolute confidence in the integrity of the operator. Now after the response by Czechia's Civil Aviation Authority, can we all be so sure? There should be no misunderstanding here, this was a potentially life-threatening incident which involved patently faulty decision making by the crew. Not only that, but the PIC with more than 20,000 flying hours was in a senior management position with the company.

It would have been a relief to learn that the regulator had taken appropriate measures to prevent a recurrence, however Civil Aviation CZ (UZPLN) released their report on the 22nd August 2019 incident <u>in the Czech language only.</u> That was on 23rd July 2020, nearly a year after the event.

The primary reason for the aviation industry having adopted a single language as the standard means of communication is safety. To ensure that information may be shared and disseminated effectively for the benefit of all. As the old maxim goes: "We should learn from the mistakes of others in aviation, because we will not live long enough to make them all ourselves!"

In a similar way this is exactly why Crew Resource Management (CRM) has been fostered, preached, taught and embedded in the operation of airliners for the past 30+ years.

What on earth were they thinking? Not the crew, that was obvious – simple negligence (and possibly criminal deviation from acceptable procedures) driven by a focus on cost saving for the airline. Plus a potentially disastrous breakdown in CRM.

But what were the people in senior roles at the Civil Aviation Authority

CZ thinking? What made them believe that they could possibly diminish the magnitude of this event by releasing a report understandable to only Czech speakers? Not only that, but the delay in its publication is significant and indicative. Seriously, this event should not have occurred in European airspace and certainly NOT with an EASA carrier. It presented a real danger of possible hull-loss with concomitant tragic loss of life. Consider just one aspect. At no time were the crew able to accurately diagnose the cause of the engine failure. In which case the possibility of *fuel contamination* could not have been eliminated.

✈

Although the final report was published in English on 27th November 2020 it leaves many important questions unanswered. The Czechia District Prosecution Office opened a criminal investigation but the report provided no details on its status. The aircraft commander retained his instructor and examiner privileges issued by the CAA. Was this appropriate? The commander was recommended for psychological evaluation by the Institute of Aviation Medicine – did this occur? If the commander was sent for psychological evaluation – should the first officer also undergo evaluation for his part in the CRM breakdown on the flight?

Accident and incident investigations in our industry are vitally important for the prevention of future occurrences. In this case, regrettably it appears the report has left more questions than answers. The link to the report is below.

http://www.avherald.com/h?article=4cbe8434&opt=0

✈

BACKSTORY: The background to this one is all over the internet. Don't get me wrong. We all make mistakes and I have made my fair share. However I have not seen many events as dramatic as this. Particularly distressing was the apparent attitude of the Czechia regulator's errant belief that they could somehow conceal the facts from the wider aviation industry by publishing the report in Czech and delaying its release for a year. Bizarre!

30 GROUNDHAPPY*

It's an oxymoron of course. We are not in the least bit happy to be on the ground. Don't get me wrong, it is always nice to finish a block of work flying airliners for a living and then spend some time enjoying days off. In our high-pressure occupation, being able to decompress is vitally important for our mental wellbeing. Time to be with the family, sort out personal admin and play with our toys. The current situation is unprecedented. Many of us have been grounded for over two months. There are factors which make that an undesirable state from several angles.

Take recency. In our strictly regulated world, as pilots we are required to regularly demonstrate our competence at flying aeroplanes. In the normal course of events an airline pilot is tested and checked in the Simulator every six months. It is called an Operator Proficiency Check (OPC).

During the two-day full-motion Simulator session we are presented with a range of emergency and non-normal scenarios. Not only is our locomotor coordination assessed, but our decision-making processes are also closely examined. I cannot think of another profession quite like it. In addition, we are required by aviation law to have experienced/recorded a minimum of three take-offs and landings inside every 90-day period. The significance of this is that by doing so, we would have practised most of the vital flying skills. All the pre-flight performance preparations/calculations and checklists, then all the prelanding process too.

In most companies however they are even more conservative. Often you will find that their operation manuals dictate that pilots should maintain a currency of three take-offs and landings within a shorter period. One of the things which has taken place early in the COVID crisis, is that both the airlines and the regulating authorities have made provisions for changes or extensions to these rules.

Common sense dictates that we should work to the regulation expiry dates. In addition, the Civil Aviation Authorities have extended the expiry dates of licence requirements by four months. This flexibility gives a method for airlines to replan their training/checking such that their pilots maintain legal currency even while sitting on the ground for an extended period. For sure, the situation is far from normal. This is the longest period without flying I have known since 1989 and it is most unsettling.

The acceptable means of compliance with the extension of validity has been achieved by the airlines setting up an online examining scheme. In my own case, my B737 rating was revalidated on the very last day of my 6 months by an online video call with a Type Rating Examiner (TRE). In a similar way to the usual simulator check, we were formed as a 'crew' Captain and First Officer for the video call.

Our application of procedures and technical proficiency was assessed by the TRE during a question and answer session. Just like the Sim it was a tense time with a pass or fail standard. Fortunately I passed the check and my rating was now valid for a further four months. Part of the reason for the online system being set-up was the fact that nearly all Simulator training centres were closed due to governmental restrictions on essential workers. Also, due to travel constraints, trainees and examiners/instructors could not travel to the centres anyway.

As a profession we are also subject to comprehensive medical scrutiny every 6 or 12 months depending upon age. Uniquely the authorities have permitted 4-month extensions where necessary for our Class One Medicals, provided we self-certify our continuing fitness to fly. Exceptional environments require innovative solutions and it is pleasing to see that the industry has moved its own goalposts to keep itself ready for business to resume. Business will be far from usual when we restart full operations. Many airlines without substantial financial reserves will fail and those that remain will struggle to remain solvent.

Recently many pilots with whom I usually operate have been asked by their companies and agencies to accept reduced terms and conditions. This is hardly surprising when you consider that the employers (the airlines themselves) have nil revenue coming in. Meanwhile they are still required to pay expensive leasing costs on their assets. It is regrettable that many pilots seem to feel 'entitled' and therefore have objected vociferously to accepting reduced salaries even when thousands of our colleagues are being fired or made redundant.

In my opinion the pilot unions have also been slow to adapt to the new reality. Some of them have been obstructive and unhelpful when companies have asked for reductions in future salaries to maintain jobs. It reminds me of the ridiculous situation in Belgium with Sabena. In 2001 the airline was facing ruin and everybody knew it. Even as the airline prepared to file for bankruptcy, the pilots walked out on strike! Unbelievably stupid. 'Turkeys voting for Christmas' comes to mind.

The maintenance organisations of the large airlines have their own challenges. Never have they had so many airframes sitting idle for so long. There are constant reports of parked airliners with birds' nests in flap tracks on the wings and in the outflow valves, APU inlets, exhaust ducts etc. The pitot tubes and sensors which were not sealed up for storage have also been found to be full of insects especially in the hotter locations where the planes have been stored.

Fortunately, the engineering teams are savvy and have responded to the new threats proactively with protective covers and regular inspections. Maintenance periodic flying has also been reduced to a minimum and this should be welcomed as it is a waste of resources and unnecessarily polluting. The threat to safety of aircraft being idle for long periods is a new phenomenon to the airlines. It should be noted it has been well understood for decades in the private jet world.

I distinctly recall in the past when we parked a VIP Boeing 757 for several weeks at a time, there were many tasks undertaken to protect the airframe. As pilots for the VIP operation we often worked closely with the engineers carrying out their storage procedures. Something as simple as closing the outflow valve on the fuselage, pulling the related circuit breaker to remove all power to the valve doors and then taping thick plastic sheeting over the outside of the hull. This prevented the ingress of any foreign bodies, insects or wildlife. Pitot and static vents were also routinely covered, tyre shrouds fitted, engine covers put in place and windshield blinds placed in the flightdeck to keep the sun out.

So much for preserving the hardware of our industry, but what about the liveware? The crews have extended downtime and many live under national quarantine restrictions. It is important for us to stay as informed and current as possible. In my case I regularly refresh myself on flying procedures (the SOPs). Whereas usually we get to practice them all for real in our flying duties most days of the week, now I find myself putting aside a few hours of each week to revise them. It will come as a shock to us all when we re-enter the cockpit after our protracted absence.

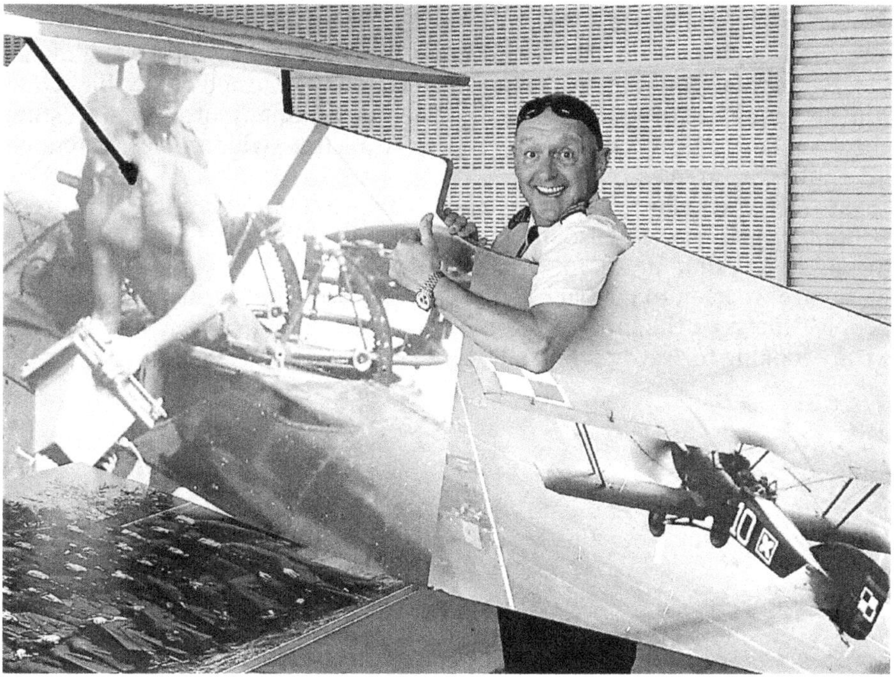

In my case I regularly refresh myself on flying procedures!

Seriously though, it is incumbent upon us to prepare as best we can. We are blessed with the internet of course. It is now so difficult to imagine how the aviation world coped before. Nowadays we receive regular updates on our charts, performance and procedures online through the company issued iPads. This gives us all confidence that we will be back in the loop very quickly when operations recommence. Interestingly, I don't hear of any of my colleagues using private Flight Simulator software to assist them with retaining currency.

I find that the most important thing to remember during this extended time of non-flying is that it is only temporary. We must not forget that the airline industry has grown steadily in every decade since Orville and Wilbur discovered the secret. Individual years have been good or bad, but it's the decades that matter. We must take the long view and stay positive. Even though there will be a slow start to airline flying, things will pick up and the cabins will be full again. The way in which we all travel may have changed forever. The prevention of cross infection from strangers will be a fact of our new way of life. Doubtless travellers and crew may be tested before flight and will be required to wear masks while in the company of others. Testing checking and certification of being COVID free will replace extended quarantine on arrival.

Let's face it, who is going to book a two-week holiday to Spain from London, if you must self-isolate for 14 days on your return? You would have to leave work for a month for a two-week vacay.

So are we keen to get back flying again? Let me put it this way. To most people, 'the sky's the limit' – to pilots, it's our home. I can say for certain... we are looking forward to going 'home'!

✈

*Groundhappy – Is an old military expression and was used to describe pilots getting itchy feet when grounded for lengthy periods.

BACKSTORY: Written in May 2020 when it was apparent that the airline industry passenger figures had collapsed. It was one bad news story after another and a very depressing time to be a pilot.

✈

31 AN EMERGENCY DESCENT

...towards the lowest cost base!

'Unprecedented' is an overused word these days. Regarding the airline industry it has been used repeatedly to describe the catastrophic reduction in public demand for airline travel. The numbers are huge and the financial stresses on the airline operators have been extreme. Whole fleets of aircraft that were once viewed as major sources of revenue have been grounded, in some cases permanently. Hundreds of airframes have made the one-way trip to the desert.

The business model of airlines has not changed much in the past 50 years. It is a high cost operation with a relatively low profit margin even in the best of times. Consider that in 2017 the average profit margin was only 9 percent. Add in the fact that it is a highly seasonal business and you get the impression that it all balances on a knife edge. But it was not always this way. Back in the glory years, the airline companies were under much less pressure to perform efficiently. It was common for the major carriers to charge what the market would bear and in so doing milk the consumer. Competition was less and profits came easier.

Deregulation of the industry in the USA in 1978 reduced barriers to entry of new carriers to the market. This increased competition and benefitted the consumer. The EU followed suit in 1993 when national obstacles to air travel were dismantled. At last the airways of Europe were opened to real competition. Monopolies were outlawed and in only three years, 20 new airlines commenced operations. Reduced fares and new routes appeared the whole process was judged a success. By 1996 charter flights accounted for more than half the market and an estimated 90 percent of all passengers were travelling on cheap fares.

All this good news for the consumer failed to alter the fact that running an airline is an expensive undertaking. Almost uniquely in the service industry sector, airlines have high capital costs, are very labour intensive and highly regulated due to strict safety requirements. In the standard airline business model the fixed costs are disproportionately high. The aircraft are expensive, crew training and employment are not cheap.

Once the routes are established the cost of adding passengers to the flights is marginal. Bear in mind that the actual passenger seat is viewed as a perishable commodity by accountants and this is a good way to look at it. Once that flight has departed those seats cannot be sold again, whether they are filled or not. In other businesses the product might be put into storage to meet later customer demand, but not with airlines. It really is 'now or never'.

Consider the fact that your competitor airline has decided to drop their price on a route that you also operate. Perhaps they have even reduced the ticket prices below cost! Your company has two choices.

1) Reduce prices to match them and lose revenue or 2) Maintain your higher price and lose even more money. As stated before the fixed costs are relatively high while variable costs are low. In addition, the customers are

price sensitive. Unlike cargo where the freight must move to market, passengers can decide not to travel if the price is too high. This in a nutshell explains the rise of the Low-Cost Carrier (LCC). Deregulation upended the high yield margins of legacy carriers that had had it so good for so long. They now faced the universal challenge of lowering their too high cost bases.

The majors were running hub and spoke operations that carry associated costs, whereas the new LCCs were operating point-to-point schedules. Add to this that the likes of easyJet and Ryanair had business models that benefited from high asset utilisation. Reduced turnround times between flights does result in significant savings in Airplane Related Operating Costs (AROC).

✈

The newcomers focused on this from the start. They trained and incentivised their employees towards higher efficiency as part of their culture. Their success could be measured by comparing Cost per Air Seat Kilometre (CASK) for all airlines. The LCCs hover around 6 cents per seat km, while all the legacy carriers are north of 8 cents, with some notably above 10.

The story of how the revenue collapsed so suddenly from March 2020 for all the airlines is well known. Perhaps not so widely appreciated are the strategies airline companies employed to counter their reversal in fortunes. Traditionally almost all airline operators make a net loss through the winter months and depend on a buoyant summer to balance the books. COVID could not have struck at a worse time.

Just when they were ramping up towards the summer schedule, with crews and aircraft all prepared, the future bookings fell off a cliff. In addition, virtually all companies had hedged* on the fuel being more expensive in the next 12 months and had agreements in place to cover that. Given that the price of oil has fallen to levels not seen since 2002, the cost is colossal.

The airlines reacted by negotiating lower pay and conditions for their employees and contractors. In some cases the salary reductions were more than 25 percent. Some companies furloughed staff and offered redundancy packages to reduce the headcount. They returned aircraft to leasing companies where possible, postponed future new deliveries and some received state support from their governments. Flight programmes were reduced by up to 90 percent.

They focussed on minimising the cash burn. Though the LCCs appeared to be ahead due to their original low-cost bases, not all will survive. Airline business models were never designed to withstand such strain. Even if consumer demand returns soon, the fact that the major European operators have been given billions of Euros in taxpayer support has skewed the market. These airlines will use these cash infusions to subsidise ticket prices and sell below cost to chase market share.

Sadly it is a race to the bottom. The prize will mean survival in a new world where airline travel is less desirable than ever. Even with social distancing and enhanced hygiene precautions, it is impossible to predict when consumer demand will return to 2019 levels. As the old saying goes, "It is easy to make a small fortune in aviation, simply invest a large one!"

✈

*Hedging - a risk management strategy aimed at offsetting losses in investments by taking an opposite position in a related asset, in this case fuel. None of the airlines could have foretold the crash in the price of oil.

BACKSTORY: Written in August 2020. This was not a good time for any of us. The airlines were all of the same mind – try to stop the cash burn. Our industry is on life-support these days.

✈

32 STABLE APPROACH

From my position sitting behind the two pilots making their approach to land, I could see that things were not going well. The B737 was well above the glideslope, not yet configured for the final descent and with way too much energy. The airspeed was eye-wateringly high and they were only just inside the limit speeds to deploy the flaps. Needless to say, as their instructor I was more than a little curious about how they would handle the situation. Surely they would announce a 'Go-around' soon, very soon. But no!

"Gear Down, Flaps 15 and landing checks", came from the captain and the first officer responded with the called-for actions.

I noted at this point that they still had the speedbrakes out and the thrust levers closed trying to get rid of all that excess speed. At only three miles on finals they were at least double the height they should have been, with 50+ knots more airspeed than the ideal. Still the engines remained at idle. They had received landing clearance and now I was seriously concerned, I was mentally wishing them to go-around, climb away safely, retract the flaps and gear, then prepare for another approach. But they continued.

We sailed over the runway threshold at around 200 feet, 30 knots fast, but at least in the landing configuration with gear down and Flap 40, still no power from the engines, the speedbrakes had been stowed. The 1000-foot point, (normal landing zone) whizzed past the cockpit windows and we were nowhere near maingear touchdown. The captain's right hand moved from the thrust levers and grasped the speedbrake lever*. I thought to myself, *'Please God no!'* But then he thought better of it and put his hand back on the idle throttles. The synthesised Auto-Radalt Calls "FIFTY... THIRTY... TWENTY!" seemed unusually loud in the flightdeck. As I looked over his shoulder, I could see we would land over halfway down the runway.

Suddenly, BANG! We were down! The mainwheels hit the runway and immediately, the captain's right hand grabbed the reversers. He pulled them through the interlock and without a pause continued pulling them all the way back as the nosegear thumped onto the centreline. The N1 gauges soared as the engines roared their response to the command of Maximum Reverse Thrust. At the same time the Autobrake Disarm caption illuminated as his feet hit the manual toe brakes for maximum wheel braking. I could see the end of the runway coming up fast now, but we were slowing. Oh yes, we were losing speed at a great rate and all three of us were thrown forward into our shoulder harness by the force of our retardation. The first officer had been virtually silent through all of this. Now he came to life as he saw the continuous red lights of the runway centreline.

"THREE HUNDRED METRES CAPTAIN!" he shouted.

I looked at the groundspeed readout and it was still ridiculously high, although rapidly reducing. I could imagine those poor brakes and the steel discs glowing red hot under the strain.

Finally we came to a shuddering stop with the antiskid system kicking in and out over the last 100 metres. As the speed registered zero the nosegear oleo bounced the cockpit back up to level. The thrust reversers were still roaring in protest. *'Ewww...'* I thought of those poor old engines re-ingesting their own exhaust gases along with all the dirt and dust from the runway threshold.

The captain and first officer just sat there in silence. We all three noted that the concrete of the runway surface ended just 30 metres in front of the flightdeck windows. At this point I pressed the 'Freeze' button on the Simulator panel next to me and the silence was total.

The captain sat very still for a moment, breathing heavily. Then he half-turned in his seat looking at the co-pilot. He swallowed and took a deep breath before he said very slowly in his deep baritone.

"I THINK... we should... have... GONE AROUND!"

The pale faced first officer nodded his agreement. I smiled behind them both in the darkness of the Simulator,

"Yes indeed Captain, I COMPLETELY AGREE with you! Gentlemen! Let's try that approach, again shall we? Please remember... you can ALWAYS Go-around!" The lesson had been learned.

✈

On another occasion, I was sitting on the flightdeck jumpseat for real. It was a night approach to land at Liverpool where I was the Base Captain for a scheduled airline. The aircraft was a B737-300. Although I was travelling on-duty I was not part of the legally nominated crew. As such, I was very conscious of the fact that it was important as an 'observer' I should make no input or verbal distraction to the crew while they performed their operational flying task. In airliner flying this is especially relevant at critical stages of flight. The first officer was Pilot Flying (PF) and the captain was carrying out the job of Pilot Monitoring (PM). It was a clear night and they had been cleared by ATC for a visual approach to land on Runway 27.

The first officer elected to disconnect the automatics and fly manually. This was a rare treat in our world of automated airliner flying. I noted that he turned in towards the final approach well above the ideal three-degree vertical profile and by the time we were on short finals (inside four nautical miles) the aircraft was nearly double the altitude for an ideal approach.

The PAPIs were clearly visible as 'Four Whites', therefore the approach was far from stabilised. As the poor captain saw this unfold, I imagined him thinking... *"Oh NO! Not with my Boss on the jumpseat!"* But he did nothing to prevent it and I sensed he was just hoping for the best. Finally, I had let it go as long as I dared. I slipped my shoulder straps and leaned forward, my head between them both.

"CAPTAIN! If you do not take control and execute a go-around from this approach, you will be suspended and probably dismissed from the company!"

My carefully measured delivery had the desired effect. Immediately he sprang into action,

"I HAVE CONTROL!" he said while moving his hands and feet onto the flying controls then, "GOING AROUND! FLAPS FIFTEEN!"

Simultaneously, he eased the nose up into a climb and advanced the thrust levers to the Go-around thrust position. I sat back quietly and breathed more easily in my seat as I slipped my harness back on again.

After an uneventful landing, he taxied in and shutdown the engines. They completed the after landing and shutdown checklists and sat in embarrassed silence. There was nothing more to say. I stowed the observer's seat away,

"Thanks for the ride gentlemen, have a pleasant evening, goodnight!" and departed the aircraft with the other passengers.

✈

Statistically between 7 and 11 factors combine to form an 'error chain' in the lead up to an aviation hull loss accident. An operating crew can break the chain by removing just one of these 'links'. Without speculating as to the cause/s of the tragedy in Karachi with flight PIA8303, on 22nd May 2020, the reportedly unstable first approach to land was likely a major lost opportunity to protect the safety of the operation.

In brief, this aircraft, an Airbus A320 was extremely high and fast on final approach for the first landing attempt. It appears that the crew lacked the situational awareness to realise a go-around was essential. They continued to the runway even though multiple warnings were evident inside the cockpit and with the ATC controller advising them they were too high.

✈

*Boeing prohibits the extension of speedbrakes, otherwise known as (Lift) 'Spoilers' while airborne with flaps greater than Flap 15 in the B737. It causes an extreme rate of descent and its use to land the aircraft would result in structural airframe damage.

BACKSTORY: One of the biggest identified Key Operational Risk Areas for European airlines in the past 20 years has been to prevent unstable approaches by pilots. In this they have been very successful. The powerful combination of active Flight Data Monitoring and follow-up with a well-run Safety Management System has been responsible.

✈

33 SALTY BARS

When I speak about 'salty bars' I am not referring to waterfront drinking establishments.

The 'bars' to which I'm referring are those which adorn the epaulettes of flightcrew. Specifically they are of the four gold bar variety. Nobody in our industry wants to wear worn looking two or three bar sets of epaulettes. This would signal that you have been occupying the right-hand seat for too long. There is a certain level of stigma attached to being passed over for command too often. To be labelled 'Career First Officer' is not a good title trust me.

But why salty? Well I heard the expression years ago in reference to a ship's captain, whose epaulettes might get that authentic aged patina from standing out on the bridge wings in all types of weather.

I distinctly recall the excitement and drama of my airline's command upgrade process when I began flying the Boeing 757 as a young first officer. A lot of 'Galley FM' revolved around which FO was next to be promoted. Not only is it a big career step to move from the right-hand seat to the left in a jet airliner, but the salary increase is similarly dramatic.

Of course it's not solely about the money, but usually the take home pay increases roughly by 50% - or at least it used to. Additionally, there is a step change in status. One month you are looked at as a JAFO (Just Another First Officer) by the cabin crew in the crewroom, and then suddenly you put four bars on for the first time. Everything is so different. You become an attractive proposition for marriage! Is this connected to your new long-term earnings potential?

I recall a story about Cathay Pacific back in the '80s. They were having difficulties attracting enough young pilots to fly their long-haul fleet out of Kai Tak. The company embarked on a large-scale programme of engaging co-pilots from the Indian sub-continent. All these new eligible young bachelors caused a big stir in the crewroom and the number of inter-crew relationships spiked...

Many of the 'Old China Hands' joked about the new *'First Officers Futures'* shares trading on the Hong Kong stock exchange. The Indians were not so gullible however and within a couple of months they had collectively been given the name *'The Bengal Lancers'*.

Airline crew love to discuss who will be the next candidate for a command course. Consequently course candidates feel tremendous pressure to succeed. I remember general euphoria breaking out among the close-knit Air 2000 crew community in Manchester back in the early '90s when one popular FO passed his course. Once Mike became a captain everyone wanted to fly with him.

I was lucky enough to do so in the week following his success and was eager to hear all about it. He recounted his experience with the Senior Training Captain on the jumpseat in the flightdeck for the final sign-off check ride. Mike operated from the left-hand seat with a real FO in the right. Airlines usually conduct the final clearance to the line in this manner.

The new skipper must then fly the first 100 hours or X number of sectors with 'experienced' first officers before they are permitted to operate with less experienced crew. This is a prudent procedure when you think about it. The last thing you want is a totally inexperienced flightdeck crew flying your jet to Palma!

They had that '50 mission' look about them

The Training Captain sits on the jumpseat on a live line flight with passengers in the back. He or she is not a member of the legally nominated crew. They are the most senior pilot present, but they are not at the controls and therefore can take no part in the flying of the airliner. They are reduced to the level of observer. It really is the very last check for the putative captain sitting in the hot seat. It is still a 'check' though and must be passed before you are able to be rostered as sole Aircraft Commander.

In Mike's case, he had Captain Allan Blake sitting behind him. Allan was very senior in the Flight Operations Department and a well-respected aviator. At the end of the flight Allan informed Mike he had passed and he would be rostered on flights as a company captain starting the very next day.

He asked Mike if he had got his uniform ready, he replied no, he thought that might bring him bad luck. Mike didn't even have any gold bars to wear on his shirt the next morning as captain. There and then Allan took off his own epaulettes and handed them over to Mike. He was still proudly wearing them when I flew with him a week later. They had that '50 mission' look about them. I bet he still has them!

One of the reasons that the gold bars look worn after a while, is that we wear five-point seat harnesses and the shoulder straps rub on the epaulettes.

Recently I lost my flightbag. It was stolen by an opportunist thief from the back of my unlocked vehicle that I left unattended for only a few minutes. The financial loss was substantial. A lot of cash which I was carrying to pay some bills personally, credit cards, phones, passport, driving licence, flying licence, some irreplaceable original documents, you name it – all gone! It was a wrench for sure. With no licence and no passport, the consequential loss of earnings meant that it was a very high price to pay for my inattention. We are talking thousands here.

But things that hurt the most were the most sentimental. I lost my Ray Ban aviators that I had bought in 1990 on one of my first ETOPS trips across the pond. Facing the future flying without them made me very sad indeed. Also gone were two sets of epaulette bars, a new set and a salty pair. You can guess which ones hurt more to lose.

Naturally I filed the police report, did my best to recover my belongings, processed the insurance claim and got on with my life. Still at the back of my mind was the knowledge that I had lost my beloved sunglasses along with two old friends – one for each shoulder of my uniform shirt. It nagged at me and lowered my mood. Imagine my joy however, when I subsequently opened the glove compartment of the car and rediscovered my lost bars!

Yes, they had been in there all the time. I had not left them in my flightbag. In my rush to go shopping on the way back from the airport, I had whipped off my stripes and popped them away in the front of the car. It was a small consolation, but a significant one. Now, I keep as little as possible in the flightbag. Fly Safe!

BACKSTORY: With my old gold bars in hand I don't plan to buy any more before I retire. Without fear of contradiction, 2020 was the worst year ever. That said, we all have to focus on the positive, if not for ourselves, then for our family and friends. I made a conscious effort to make the best of a bad situation after losing all my stuff. My fiancée and I went sailing round the Greek Islands for the whole month of September, I loved it. What else could I do? Without a passport I couldn't leave the country and with no flying licence or medical I couldn't even leave the ground!

34 MINIMUMS

The winter of 1987/88 was quite an adventure in my aviation career. At the time I was flying for Grosvenor Aviation as an Air Taxi pilot. I had left the military in mid-1987 and the only piston engined twin I had ever encountered was the GA-7 Cougar we used for the Instrument Rating test. That took all of five hours, so it hardly counted. At Grosvenor (nothing to do with the Earl) we operated two Partenavias and three Piper Navajo Chieftains from the southside hangars at Manchester International. It's all gone now of course, after the demolition of the southside and construction of the second runway.

Back in the day we used to be quite busy at times. One of our regular money earners was the nightly three-legged mail flight with the Chieftain from Manchester. We flew 'GRV015 Alpha' empty from Manchester to East Midlands Airport, then '15 Bravo' to Belfast carrying the mail and finally '15 Charlie' ferrying empty back into Manchester in the early hours of the morning.

We departed Manchester empty in the evening and ferried into East Mids. I'll tell you how long ago this was... the approach controller was callsign 'Castle-Don Approach', the airport being situated so close to Castle Donington of course.

On a clear night it was a shame to take the money. Seriously it was a nice short ferry flight there, then taxi onto the cargo ramp to wait for the Royal Mail crews to load the aircraft. I say 'we', but really it was all solo. In fact, the aircraft was flown with the other pilot seat removed to reduce the weight. The engineers fitted a removable blank wall covering the rudder pedals on the right-hand side of the cockpit to prevent any control obstructions there. So we really were *single crew,* which in those days was legal provided we had an autopilot fitted.

The PA31-350 was a great performer when it was empty. When loaded up to max takeoff weight - not so much! In retrospect we were fortunate to have achieved all that we did with that mail contract, because the safety margin was wafer thin on every heavy takeoff out of EMA. And of course, those departures were always late at night. Around midnight, when you were tired, sometimes in bad weather. It makes me shudder to think about it these days.

Another operator was not so lucky. There was a fatal accident on the night of 25th January 1988 at East Mids with a mail-plane. That night my colleague Malcolm Kavanagh was operating our "GRV015" series of flights.

According to reports, the King Air 90 from National Airways based in Southend crashed during a Go-around from an instrument approach to Runway 09. On the ramp waiting for the loaders, Malcolm heard the explosion which then lit up the sky. He came home empty to Manchester that night as none of the mail-planes operated.

"I was there on the ramp parked facing West. I was filling in the tech log

when suddenly the whole sky to the West lit up and it was bright enough to light the inside of the aircraft. Very shortly after the fire engine drove past the ramp toward the flames. The weather was bad with fog and low viz during my approach. I was fully prepared to go around but at the last second saw the lights. I think I was the last to land that night."

Captain Malcolm Kavanagh – aviation professional

Much later, reading the AAIB report, it was clear that the weather was right on the limits for a CAT 1 approach to minimums. The height of the aircraft at Decision Altitude is only 200 feet above the runway. Now, think about it; 200 feet above ground, in cloud, flying on instruments, doing 100 plus mph... that's a very risky place to be! You must make your decision quickly when the time comes. And as single pilot, there is nobody to help you with decision making.

Most sobering from the accident report was to read the pilot's transmissions on the Tower frequency.

"National Six Four Six, got to Go-around" and then ten seconds later, *"Crashing... National."*

We don't know what occurred in those ten seconds. However the aircraft impacted the ground from a 45-degree dive. The investigators determined this likely resulted from a stall.

Significantly, the landing gear was still extended at impact. There are a couple of possible explanations. Firstly, it could have been a simple omission on the go-around manoeuvre and contributed to the lack of climb performance as the aircraft stalled.

Alternatively, it might have been that the pilot was still looking to pull off a landing, even though he called *"got to go-around"* on the radio. Ten seconds is a long time during the go-around. By that time the machine should have been well established in the climb to a safe altitude. Something went wrong and the pilot knew it when he called *"crashing."*

Finally the words he used to indicate he was aborting the approach. *"Got to go-around..."* struck me as odd. It certainly is not the call I would make in the same situation. For me it would have been *"GOING AROUND!"* and I would have almost shouted it.

I was blessed to have received my instrument training in the most disciplined flying regime of all – the military. As a result I have firm opinions on things like "Decision Altitudes." Instrument approach ***Minima*** are exactly that, the LOWEST limits that we can work to, they are non-negotiable. While just a number on an altimeter it's there to protect us and everyone else involved.

It is hard to discuss this without feeling sorrow and regret for the pilot, plus his family and friends. It may be that he did not make a mistake that night, perhaps he was just unlucky. However it is more likely that the accident was a result of pilot error in the execution of an all engines operating go-around.

I am not one to judge, because I have been there. That very special place. The place where you must DECIDE... close to the ground, still in cloud, travelling at over 100 mph. You can't see Jack____ and your altimeter is approaching minimums. Yes, I recall clearly, even now more than 30 years later. I was THERE, that winter on more than one occasion going into Aldergrove, fully loaded with mail. The METAR having indicated that I would be right down to minimums before I saw the runway. One night I was making the ILS approach to runway 25 in just such conditions.

I prepared the cockpit well for the approach and diligently ran through all my checks ahead of time. One of the big advantages with the Piper Chieftain was its autopilot.

It had a Flight Director and would fly a coupled ILS approach accurately. I always used this function on s***** nights and for sure this was one of those! I set the instrument lights to the correct brightness, they were clear, but not too bright. I switched OFF the landing lights to increase my chances of seeing the runway/approach lights at Decision.

"Grosvenor One Five Bravo, ready for the approach" I called, then I received the final vectors for the localiser.

The Chieftain locked onto the ILS localiser beam like a champ and then I waited for the glideslope. Here it comes and sure enough, the autopilot captured the glideslope and lowered the nose. All I had to do was to bring back the power setting and deploy flaps and gear for landing. I watched the instruments like a hawk, cross-checking and rechecking the altimeter setting to be sure I had the correct QNH pressure set. The Outer Marker came and went, then shortly afterwards I noted the altimeter passing 1000 feet.

"Two Reds, Two Blues, Three Greens, Full Flap..." I had the habit of talking through the checks on the intercom in my headset. Sometimes I talked to the aircraft too. With my final confirmation of the levers all being fully set in the correct position, I was ready.

I could see the fog getting brighter and brighter outside the windows in those last few seconds. We were tantalizingly close. Then suddenly, decision? NOTHING SEEN! GOING-AROUND! Simultaneously, disconnect the autopilot, smoothly apply full power, positive rate of climb, gear up, lock on to the attitude indicator and hold that pitch! Trim as required and as soon as she's accelerating, flaps retract, YES, we're climbing away now.

"Grosvenor One Five Bravo Going-Around" Air Traffic immediately wanted to know my intentions. There was no doubt in my mind,

"Diversion to Prestwick!" so they cleared me to climb on track towards Scotland.

The rest of the flight was uneventful. I arrived in Prestwick, refuelled and waited for the weather in Northern Ireland to improve before returning there to deliver the mail. When I finally disembarked in Manchester later the next morning I was exhausted, but pleased I had made the right decision.

✈

35 FLYING THE PM

There are many VIP flights going on in the world. The availability of private jet travel for the rich and famous has never been greater. You only need two things. A reason to travel and money - lots of it. While it is hugely expensive, private jet travel does provide a unique experience for the cash. With enough money or influence, you can rent, buy or borrow a jet.

"I love to travel in a private jet," Sir Michael Caine once said, "if you have a private jet, I will be your friend." Naturally the loan of a jet to someone else is sometimes used as a cover up for what are called 'grey charters'* but this is rare. The vast majority of "Here; take my jet," offers are genuine. It is like borrowing a friend's car, but on another level.

The private jet industry services the need for individuals who require high levels of security and assured distance from other passengers. Even in the time of COVID demand has continued. Whether it is short hops in chartered helicopters or long-haul flights in large cabin bizjets, it is still there.

Moreover in some areas the market has been thriving. Well before COVID, I flew my share of VIP and VVIP passengers. There is a difference, between VIP and VVIP. Head of State flights, and Royal flights come under the heading of Very, Very, Important.

✈

The charter broker sounded ill at ease on the phone.

"Well it's a tricky situation James. Number 10 really want this flight – both sectors in fact - but the Italians are not cooperating." He continued, "they won't give you fuel on arrival at the military base, they say they have used all NATO's jet fuel in Southern Italy."

I was trying to visualise all the possible angles. Already I had asked questions which the broker could not answer. It was important to get these flights right.

"Perhaps I should speak directly to the Prime Minister's PA? That way, we can be sure what they need."

Initially he was not keen on the idea. It is against the Charter Broker's 'code' to allow direct contact between client and operator. They worry they will get cut out of the deal. I assured him that we would respect the integrity of the client-broker relationship. We had worked together before, so he knew I could be trusted and he gave me her mobile number.

I called and explained who I was and confirmed what I needed to know. She replied,

"Well the Prime Minister wants to fly direct from Heathrow, spend a couple of hours on the ground at the Italian Air Force base, meeting with UK RAF personnel based there. With the Press of course. And then to fly onward to Karachi."

I assured her we would investigate and confirm the feasibility. Next I called another UK mobile number I had been given by the broker. This one was supposed to be the senior RAF Officer at the Italian Airbase – it rang for ages before he answered. I introduced myself and verified who I was. He also confirmed there would be no fuel available for us, but also something else,

"They are operating aircraft from the NATO runway now because the main runway is closed for resurfacing. The emergency runway is only 30 metres wide. Normally this is the taxiway, can you operate the 757 on there?"

"I will check and get back to you.** The runway width is good for the seven five, but I need to see the Pavement Classification Number to be sure the surface will bear the weight, also the manoeuvring area."

I contacted our Ops guys, but they had no answers,

"Captain, it's up to you really. You're more familiar with the performance of the 757", which left the ball in my court. I studied the aerial layout of the airfield and could see more problems for us. The NATO runway was wide enough for landing and takeoff, but where to turn the aircraft round? Then I saw at the northern end of the runway there was a large concrete ramp for fighters ready to launch.

In the UK this would be termed an ARP (Aircraft Readiness Platform) so that a military formation of fighters could hold together very close to the threshold. Measuring the entrance and exit carefully, this would suffice, and we could do a 180-degree turn, no problem. The maingear track needs 36.6 metres width and the area was 45 metres wide. It was tight but doable.

I returned to the questions of the fuel and the landing weight. To reach Karachi would need plenty of fuel. It was possible to carry enough fuel out of London to go all the way, but that would mean we would land heavy into the airbase. Our landing weight with all the gas onboard to fly to Pakistan put us way over the PCN of the runway and taxiways. None of them could bear the weight of a heavily laden B757.

I made more calculations. We really did need to refuel. It was time to call the RAF Squadron Leader again. The Italians would have to make a choice. Either give us fuel, or let us land overweight on the runway. This time he picked up straight away and I explained the situation.

"Please give our compliments to the airbase commander and inform him that he must decide. Either he provides fuel for us or he grants authorisation for a one-off, landing at a weight exceeding the PCN of the runway. We do need an answer and we need it quickly, this is a Prime Ministerial flight. Perhaps I can send it to you in an email?" But it seemed that was not going to be possible,

"I am sorry, we will not correspond by email as it is insecure. However I will inform the Station Commander and revert!"

Well revert he did and guess what? They still had some jet fuel left in southern Italy! It seemed that 'Harry Staysh'*** didn't fancy our big old bird tearing holes in his only runway to launch his fighters from. So he agreed to refuel us for our trip to Pakistan. This was all to the good. Our landing weight would now be very light for a 757.

"if you have a private jet, I will be your friend."

Which is how, only a week later we came to be sitting in the flightdeck of the VIP B757. Waiting and waiting, parked on the ramp at Heathrow, outside the VIP Terminal waiting for the Prime Minister to arrive. We had all our passengers onboard now except for the last few which included the PM himself. We certainly had the journalists onboard. I recognised a couple of their names from the media.

They were seated far to the back of the VIP cabin. I was sure they were okay with that. In the 757 the bedroom is in the middle, which makes the forward few rows of seats more private. No doubt the VVIPs would be happy with that. This Boeing 757 had been beautifully transformed into a luxurious private jet, with only 62 first class, white leather seats and a double bedroom. In fact, not just a bedroom, but an adjoining en-suite private bathroom too.

B757 Bedroom – a very comfortable way to travel

Few aircraft in the world combined such exceptional large aircraft performance with superb comfort for large VIP groups and it was available for charter.

Our crew were all hand-picked and vetted for this mission. A third pilot – another captain – accompanied us to ensure we could achieve whatever duty time was required to complete the flights. Pakistan is more than 3300 nautical miles from Italy and it would take all night to get there. The first officer had prepared the cockpit. We were briefed and ready.

My first sector would be as Pilot Flying to the secret military airbase in southern Italy. We had discussed the challenging landing procedures. The runway was long enough but only 30 metres wide. There was Rotary Hydraulic Arrester Gear at both ends of the runway. The RHAG cables are to stop runaway fighters from over-running. We could trample these cables at low speed in the 757. But if we hit them at high speed it would cause severe damage to our landing gear.

Moreover the final approach would have to be visually judged and manually flown. This was not going to be an easy flight and I could feel the tension in the flightdeck. We could not afford to get this wrong with the Prime Minister and senior government officials onboard. We had a couple of very senior officers from the Ministry of Defence as passengers, this was no ordinary VIP flight. But now, we were running late. *Where ARE they?* The first officer looked across at me and said,

"Well, we've missed our slot..." I nodded,

"Yes, I know. Don't worry about it. ATC knows we're here and they know why. Traffic may be bad coming out of central London."

Suddenly, there was movement at the doors of the VIP terminal. Some of the suits came out, the ones with the earpieces and bulging jackets. They were looking around them. The Prime Minister and his entourage came striding purposefully towards us.

I instructed the FO,

"Get the clearance please" and the he immediately called Heathrow Delivery, wrote down our departure clearance and read it back. Now the Senior Cabin Crew confirmed,

"All passengers onboard Captain, may I close the doors?"

"Yes, go ahead, we will be on our way directly."

The FO looked across at me, alarmed I might be raising expectations too early. *After all we missed our ATC slot half an hour ago.*

"Call for push and start please," I said with more confidence than I felt.

When Air Traffic did not respond immediately, he shrugged his shoulders as if to say, *'well, I did tell you captain...'* A few seconds later they came back with our callsign.

"Cleared push and start, call ready for taxi."

He looked at me like I was some sort of magician! But it was logic not magic. Usually we are treated like any other private jet, occasionally we have some priority depending on the pax and our airport of departure. Today we were departing from London Heathrow and our guests onboard included the heads of government. Our priority status could hardly have been better.

Behind us in the galley and forward cabin I heard the crew securing for takeoff and delivering the safety PA. Then they called us on the intercom to confirm Cabin Secure. As we got the wave-off from the pushback crew-chief and the tug reversed away we saw a long line of planes waiting for departure. We were instructed to take a parallel taxiway towards the departure runway threshold. It became clear we were to overtake some twenty airliners in the queue!

"Seat the cabin crew please. Before Takeoff Checklist."

I smiled as we overtook mostly British Airways aircraft. *Here we are going to the front of the line with a boring callsign, in a private 757 with no visible markings.* Then we heard the tower controller with our callsign, "Surface wind 280 at 10 knots, clear for takeoff runway 27 Left" I glanced at the clock – only nine minutes after pushback, nice. As we crossed the nose of the first in line the controller said "Speedbird 123. AFTER the departing 757, line up and wait 27 Left." By now we were all smiling as we started the takeoff roll...

When we were established in the cruise, I passed over control to the first officer and got on the Satphone. We could now relay an accurate ETA to the RAF Squadron Leader. I got through on the first attempt and he was delighted I had called. He had news.

"Pass the message to your guests that a private jet from Rome landed here ten minutes ago. The head of the Italian Air Force AND the government minister for aviation arrived. ATC are expecting you. The taxiway will be clear for you to manoeuvre on the Quebec ramp."

I wrote down the details and left the other two pilots in the cockpit. After adjusting my tie in the mirror of the flightdeck door I entered the forward galley and attracted the attention of the Prime Minister's PA through a gap in the curtain. She stood up and came into the galley where I told her that the top-ranking officials were waiting on the ground for us. She asked if we had received any indication that Signore Berlusconi the Italian Prime Minister might be there?

"Well Ma'am, I don't think so," I replied seriously. "Nobody mentioned any clowns!" She laughed and returned to brief the PM.

Shortly afterwards a request came for one of the ministerial aides to use the Satphone. He wanted to speak directly to the Senior RAF Officer, so I dialled the number and handed him the phone. Those of us who overheard this end of the call were amused to find out that Number 10 was concerned about which RAF people would attend the Press Conference. There had been some recent redundancy notices sent out to active serving personnel and they might wish to ask embarrassing questions in front of the media! There was a war going on, yet somebody in high command had decided to give notice of redundancy to front line troops. You could not make it up.

The aide went back into the cabin and we prepared for descent – not long now. The weather was clear. We would make our approach to land on the northerly runway with very light winds and confident that we could depart later from the southerly runway. The high-pressure weather was set fair and the wind was dropping for the night.

The controller brought us round for a final approach from eight miles out and we flew an almost continuous descent. I noted that we flew past Taranto harbour on our right-hand side. The PAPIs stood out clearly, but that runway did not look wide enough. I was glad we had measured the width precisely on the satellite images. I ensured we would land beyond the RHAG cables and it was pleasing to land lightweight, into an unknown airfield with a narrow runway. We had such a very slow approach speed; the Vref was below 120 knots on finals. The B757 is such a great performer in every way, power, brakes, flying speeds.

And then we were down. The autobrakes kicked in and we didn't need more than idle reverse. I disarmed the autobrakes and we retracted the flaps, completed the after-landing checks. We still had quite some way to taxi before we would be on the ramp. It was exhilarating to be somewhere totally new and I manoeuvred the aircraft carefully through the narrow Bravo exit to the Quebec taxiway. *Hmmm, 45 metres wide?* It must be.

I slowly eased the tiller over to the right and lined up where I thought the starboard maingear was, within a few feet, got the speed down to just below 10 knots, then started the turn. I wound on all the tiller to the left and increased the power on the outboard engine. The aircraft responded well. I was glad we were so light as we turned in a very short arc. *As tight as a London taxicab,* I thought to myself as we made it with apparently room to spare.

Then we backtracked the runway, taxiing very slowly over the RHAG cable again with a big bump. We vacated onto Oscar ramp and shutdown with the marshaller. They had said this ramp could take two C130s, but that would have been a squeeze. For sure, our big-winged airliner filled all the space.

✈

Few aircraft in the world... superb comfort for large VIP groups

The PM and his entourage left the aircraft and the cabin crew went straight to work, cleaning and setting up serving equipment for the next leg and preparing the bedroom for the Prime Minister. This flight would be the long one, eight hours to Karachi. Meanwhile we pilots and the travelling engineer completed the turnround. The refuelling was not altogether straight forward. The military fuel truck driver had not refuelled an aircraft like ours before and needed our assistance.

We ensured that the flightplan had been filed and we discussed the departure from the southerly runway. Again the RHAG had to be trampled at low speed, which meant a shorter takeoff run for us. We prepped for a full power takeoff firstly due to the reduced takeoff distance and secondly due to our security concerns. We were in southern Italy with the heads of UK government onboard and departing from a military base while NATO was conducting bombing raids against another country not very far away. While our arrival may have been a secret, it was obvious we were here now. A large white airliner at takeoff or landing makes an easy target for an individual with evil intent.

We were not worried about the possibility of a missile attack, this was not Baghdad or Mombasa . We were certain this would be beyond the capabilities of any terrorist group in Europe at that time. However a terrorist group could mount a small arms fire attack on a slow-moving target, close to the ground.

The risk prompted us to go completely dark as we got airborne. When I say dark, think absolutely blacked-out. You know that announcement they make in airline flights; “It is normal to dim the lights in the cabin during hours of darkness.” We took it a stage further. I instructed the cabin crew to switch all the lights off completely at the start of our takeoff roll. In the flightdeck, I briefed my colleagues that as we rotated, Pilot Monitoring would switch off ALL external lights. Every single one. By that stage we would not require outside visual references as I would be flying on instruments.

Overly cautious? Maybe. To be fair, there had been no recent attacks on airliners carried out in this fashion, but I knew it was possible. In aviation we must constantly think about TEM – Threat and Error Management. At an early stage of my airline career it was instilled into me, that often the captain of the aircraft is the final arbiter of flight safety. We are the final link in the chain and over the years I have seen many occasions where an accident or incident was actively prevented by the pilot in command. On the other hand, the pages of aviation history books are littered with accidents caused by the error chain remaining unbroken all the way to the end.

Multiple factors contributed to my decision. We were transporting the Prime Minister and other cabinet members during a period when our country, was flying bombing missions against an unstable dictatorship with a long history of sponsoring terrorism. Furthermore we were in a foreign country at a military airbase, the security of which we could not properly assess. This was serious business It all had a vulnerable feel to it.

Now the aircraft was fully ready again. We were just waiting for another VVIP, the most senior politician in the British government. We were pleased when we spotted journalists heading our way – the PM would soon follow. Only minutes later and they exited the building under the arc lamps walking across the ramp in the gathering gloom. The Prime Minister walked with a confident air of authority flanked by senior Italians, including the base commander. I instructed the crew,

“As soon as he’s onboard, call for startup and call Air Traffic to activate the flightplan.” I then buttoned my jacket and joined the Senior of the cabin at the L2 door.

"Welcome back Prime Minister" I smiled.

"Thank you captain" he replied as he shook my hand while crossing the threshold. I checked with his PA that all were back onboard, then turned to the Senior,

"Clear to close up. Let's go."

B757 bedroom with ensuite bathroom

There was quite a party atmosphere developing in the forward cabin as I made my way to get past them all. The PM was there jacket off now, smiling broadly as he was discussing the bedroom with General _______ (Head of the Armed Forces). He was standing there looking very smart in his full uniform and had been suggesting that there might be some question regarding who would be occupying the bedroom. The PM was having none of it as he said in his poshest English tones with a twinkle in his eye, enjoying the moment.

"RHIP* Johnny I'm afraid! The bedroom is mine," and then laughed out loud along with everybody else. I was delighted, it is rare that you get to see a general put in his place and there are not many people who could do that.

I was grinning like a Cheshire Cat as I came back into the flightdeck. The others asked,

"What's so funny?" I shook my head,

"Tell ya later. Pre-flight checklist please."

I slipped into my seat, moved it forward and was clicking the harness straps in as the FO started reading and I was responding. This part felt good.

All the planning and briefing was behind us now. We have come to the fun part... *Man, let's fly!*

"Commandante! You are cleared for the engines start. Two then one." Then shortly after I tell him,

"Grazie Mille my friend. Two good engines start. Clear to disconnect, we will see you ahead of the aircraft. Arrivederci!"

There was a click as he pulled out the jack plug. Then we saw him in his camouflaged military fatigues and Hi-Vis waistcoat walking out front. With Flaps 5 selected, checks complete,

"Call for taxi please." Now we were on our way.

We backtracked the northerly runway slowly, running our final checks and waiting for Cabin Secure. Again, we had to vacate and manoeuvre on the ARP, fortunately the 757 is blessed with great lighting for just such an occasion. As we were cleared for takeoff, I made the PA to the cabin myself regarding all the lights going off. There would be no misunderstanding here,

"Ladies and gentlemen, it is normal in the hours of darkness for the cabin lights to be dimmed, you may continue reading with the reading light in the panel above your head. Thank you."

In the cockpit, we replied, "Clear Takeoff" and I pushed the thrust levers to stabilise at 1.10 EPR, the FO called "Stabilised" and I pushed the EPR button on the MCP. The autothrottle moved the levers forward, I followed up with my right-hand noting the EPR gauges achieving rated thrust. I was also aware of the fabulous roar from behind us as the RB211s gave their all. Even with 30 tonnes of fuel onboard that takeoff was a rush!

The runway edges blurred as we went through "Eighty Knots, Vee One, Rotate!" so quickly and I applied back pressure to the control column. "LIGHTS OFF!" and there was a series of rapid clicks as PM hit the switches. I glanced outside... total darkness. Like a homesick black angel, we soared towards the stars. We cleaned up and were handed over to Apulia Control. We said goodbye to our military friend. Passing 10,000 feet we put all the lights on again and released the crew to commence their service in the cabin. At the same time, we reduced the climb power to something more economical. The danger was passed now.

As we reached our cruising altitude, there was palpable relief. The thrust levers slowly retarded and the noise level decreased Simultaneously to a low rumble. The three of us coordinated our rest periods and crosschecked our ETA for Karachi. With just seven hours to go, I swapped with the relief captain and went aft.

In the cabin I could see the crew were serving pre-dinner drinks and canapes. I moved through heading for a little quiet area we had screened off for crew-rest. There were blankets and pillows prepared ready. I set the alarm on my phone and snuggled down for a couple of hours. When I awoke, the cabin was now in darkness. Dinner had been served and all was quiet.

There was hardly any movement among the passengers. They were all either sleeping, reading or watching movies.

I returned to the cockpit and got an update on our progress. All good. We were still boring a hole through the night sky at 500 mph, heading east. Soon the dawn would come racing towards us at 1500 mph, not a pleasant thought.

Only a few hours later it seemed we commenced our descent and approach into Karachi. I was now in the right-hand seat doing the Pilot Monitoring job and the FO had returned to sit on the jumpseat. He looked much better after his four hours of rest. We got some very lazy radar vectoring over most of the urban area during the initial approach towards the ILS, which made us think our arrival was being sponsored by the Pakistan Tourist Board.

As we touched down you could smell the city coming through the air conditioning system as the cabin fully depressurised. It was just after lunchtime. Hot, humid, smoky and exotic - with a hint of Asian cooking as a background flavour. We were escorted and led by a follow-me truck to the VIP terminal.

Here they had a reception committee and a military guard of honour waiting. The marshaller guided us so the L2 door would be precisely abeam the end of the red carpet, very nice. As soon as we had finished the shutdown checklist, I moved back through the forward cabin to be at the L2 door for disembarkation.

The senior was already there and she looked calm and composed, which was reassuring. The groundcrew secured the external steps to the aircraft and there was a double knock on the door. She looked at me, I nodded. As she opened the door, the full heat of the afternoon hit us.

At the foot of the steps I could see the reception committee was all there ready, including a few westerners in suits. I presumed they would be from the British Embassy. The PA looked over my shoulder and spotted who they were and we stood back then for the Prime Minister to disembark first as had been arranged. I heard her say to him very clearly in a loud whisper,

"Bottom of the steps, PINK TIE! British Ambassador" and then he was off. Very clever really, because the PM would arrive onto Pakistani soil and be able to say directly,

"Good afternoon Mister Ambassador!" Meanwhile the 'Ambo' would be thinking, *Gosh, he remembers me!* I smiled at the thought.

Soon all our guests had departed, including the media folks who left via the back door – *tradesman's entrance for you lads!* We commenced cabin cleaning. There was much to do. We would not be back onboard the aircraft for the next two days. Meantime the pilots assisted where we could, particularly in bringing spare catering equipment up from the cargo holds. Finally, three hours later our mission was complete and we were taken to the crew hotel in an escorted crewbus. Fortunately it was a convivial environment for the next 48 hours as we were not allowed to leave the building.

Two days later and it was time to return. This would be a night departure direct to London Heathrow, so we arrange to arrive at the jet two hours before we are due to go. There is always so much to do. In addition, something is bound to go wrong and we will need to fix it. There is a continual element of low-level stress operating VIP flights with only one aircraft. If we have a technical issue, it is down to us to get it fixed ASAP, even if (God forbid) we have passengers onboard at the time. That really is the worst scenario.

Maybe there is trouble starting one of the engines, as we experienced in Africa with a President onboard... *Oh My God! I was hoping the cockpit floor would open up and swallow me whole!* All I could say was "Seriously!?" With some jiggling and a prayer, we got it running at the second attempt, but that was a moment.

Eventually free from the airport security people who were on some sort of go-slow, we arrived on the ramp at the 757. Our first task was to check all the numbered seals were intact. These we had put on every hatch, door and opening of the hull. If they were unbroken then we could be sure nobody had been inside the ship since we left her.

It takes a few minutes but it is worthwhile. The alternative is for the crew to perform a complete security search of every single part of the airframe – no joke, it is a real PITA. With a sigh of relief I confirm that all the seals are where they should be untouched. Now let's get on with it.

We call for the refueller and the catering. Our travelling engineer (called the 'flying spanner') completes all the pre-flight maintenance checks, while we three pilots go through the flight documents. There is the flightplan, pages of weather charts across half the planet, NOTAMs and overflight clearances.

All have to be checked through before we can decide upon the fuel load. I can see by the forecast on our route that we are lucky it should be a smooth flight. I cannot tell you how annoying it is when there is turbulence on a VIP flight. I know I shouldn't think it is any different, but somehow you just want to deliver the perfect customer experience. Somebody is paying handsomely for the privilege of flying on our beautiful VIP machine – in this case it is the British taxpayer. You do want them to feel they got value for money.

Now it is approaching one hour to go and our media passengers arrive. The rear cabin is ready for them, so they go straight onboard. As I look at the faces of the journalists, there are several famous ones among them. Not surprising really, this is a big story. As I pass through the cabin a few minutes later to check on the rear galley, I can see they have already made themselves at home. They are sprawled around our club four seats which face each other and they are playing cards. Also I see they have cans of beer open on the card table – they look happy. I smiled and said to the party in general,

"Bravo chaps! Make yourselves at home. I see you've found the Carlsberg then!" to which one of the older hacks joked back in theatrical tones,

"We are known as... *The Sniffer-Outers!"* General amusement all round.

Privately I knew that they were not freeloading. Their news organisations were paying heavily to send these senior journos on these flights accompanying the PM. They were paying directly to the British Government! Very smart. The price per seat was eye-watering and it obviously helped to subsidise the trip for HMG. No doubt the taxpayers would be pleased if they had known.

Outside now and I am making sure that the baggage loading is going well. From experience I am aware that all sorts of things can go wrong here. We always made sure to supervise activity with the loaders, even though, as in this case, the handling agent at the airport sent their own security people. The dispatch supervisor approaches me looking worried. I mentally prepare myself. *Uh-oh! Here we go. There must be a problem.* He says,

"Oh Captain. We have a difficult issue. Our Prime Minister has given a very important gift to your Prime Minister. It is a large wooden dining table and chairs... may we be putting it in the cargo hold Captain?" I thought only for a moment. Phew! Easy one! And replied,

"Yes, no problem, we will take it. Bring it out to the aircraft." In my mind, I simply decided that we would get it into the hold, no drama. I was determined that regardless what size it was, it would not delay us. I could imagine the stress of trying to explain to the PM why we were waiting to close the doors "Because we've got your dining table stuck in the cargo hold doorway!" I mused, *as a last resort we could cut the legs off the damn thing if we had to...*

VVIP Flights are certainly different

Then soon enough, we are all ready. Waiting again. Which is fine with us.

Our major concern is to be ready in advance for the customer. Ready to close the doors and start engines as soon as the last passenger has boarded – usually the principal. Soon we have exactly that situation as the PM climbs the red carpeted steps and the cool conditioned air of the cabin hits him.

I wave to the ground party as we close the door and got forward again into the cockpit. The boys are ready here too and calling for pushback and start from the VIP terminal. With engines running and a flightplan of eight and half hours to London, we are soon taxiing to the runway. Nil logo lights and with the same brief again, we are planning for our invisibility cloak departure on rotation. The cabin crew have the same message. This is not a place to take any chances.

This time I am Pilot Monitoring and as I call "Rotate", my PF colleague calls, "Lights OFF!" I flick the switches and outside goes completely black. The tower controller must have been watching us closely, because he transmitted, addressing our callsign,

"________ *All your lights have failed!"* In the darkened flightdeck we all laugh and I simply transmit cheerfully,

"Roger!" The climbout is steep as we keep the speed back, letting the autopilot maintain a high pitch attitude, then once we pass 10,000 feet we relax. Seatbelt signs off, lights back on and the cabin angle lowers as we accelerate towards home. The homebound flight is mercifully smooth and uneventful. The cabin crew performed their usual magic with the catering and the passengers seem very happy.

The sole point of interest occurred when we were in the last half an hour of the cruise. It was now the early hours of the morning in the UK. We were still working with Maastricht ATC and they said,

"You will be number one for landing. Now call London Control on frequency..." Which was something I did not expect. How would Maastricht KNOW our landing sequence for Heathrow? Hmmm... Very odd. As we called London it became clear. WE WERE the only arrival! Of course, the night jet ban. There would be no other flight arriving or departing until 0600 local time.

It felt completely surreal as we ghosted in over a sleepy capital city at 2am. We were given a continuous descent all the way to Heathrow's runway 09R. My co-captain managed to pull off a nice greaser of a landing, with minimal braking and idle reverse thrust. Clearly our flying spanner was impressed.

"Like a cat pissing on velvet!" came from the jumpseat and I grinned.

"Good job Gary," I said in appreciation of his landing then, "have you thought about taking it up professionally?" To which Gary chuckled and the spanner guffawed. He had not heard that one before.

Once on stand at the VIP Suite the doors were opened. The principal and senior members left the aircraft first. They were grateful for a nice flight home, although everyone looked worn out. I went down below to where they were unloading the cargo holds. I was interested to see if the dining table and chairs had made it. I should not have worried. There, on one of the baggage trucks was the furniture – it was all flat-packed and bubble wrapped, just like you get from Ikea! *So glad I didn't stress about that one.*

With the last of the passengers and baggage away, the 757 was towed over to a nearby cargo parking area on the south side. Now we were cleaning again. Tomorrow we would be able to ferry the aircraft out. For now we needed to rest and eat some five-star leftovers. It had been a long mission.

✈

*Grey Charters – flights defined as private but flown for commercial gain behind the scenes. Illegal flights basically.

**I knew from operating into places like Zakinthos in the early 1990s that 30-metre width is possible for the B757. ZTH had poor approach and runway lighting, plus is was not overly long. We used to go in there at night with charter flights from the UK. It was always a Captain's landing.

***Harry Staysh – UK Military (RAF) slang for Station Commander

**** June 2014 – Peshawar.

RHIP – Rank Has Its Privileges

Peshawar Terrorist attack - 2014

Flight PK756 from Riyadh in Saudi Arabia had 178 passengers on board. It was hit by a volley of machine gun rounds as it came into land at Peshawar. Police said one bullet narrowly missed the captain. A Pakistani woman passenger was fatally wounded when shots penetrated the cabin.

In addition to the woman passenger who died from a head injury when a bullet struck her, two of the cabin crew were also injured and taken to hospital after landing. According to news sources a sub-machine gun was used in the attack. Of the 12 bullets fired, 6 struck the aircraft. The gunman was tried and executed in October 2017.

Also, in June 2014. Terrorists attacked Karachi International Airport itself with machine guns, grenades, and explosives. There were some 38 deaths as a result including all of the attackers. I read the report with sadness some years after we had flown in there.

✈

36 THE BIG PICTURE

How often do we hear 'You're too close to the problem' or 'You can't see the wood for the trees? The two expressions have identical meanings. Someone thinks you are missing 'The Big Picture'.

I heard it once when things had gone badly wrong out on the line. We had a tech aircraft and I had called the Operations Controller for support. I suggested a couple of ways we could replace the aircraft, with another.

The Controller was in no mood to listen.

"That's all very well Captain, but you don't have the Big Picture"

Well - I could have spat my dummy, but instead I tried yoga breathing. Inhale slowly through the nostrils and gently out through the mouth.

I managed a cordial goodbye without losing it. I remember thinking that next time I would be less keen to offer my suggestions.

✈

Which is why I was reluctant to call Ops when we had a problem down in Catania, Sicily a few years ago. It was early in the morning and we were planning to fly 4 sectors. Firstly Rome Fiumicino and back, then Turin and back. The crew was based in Catania. The weather forecast showed Fiumi was fine, but Turin was forecast fog all day. The TAF was showing 200 metres visibility for the entire period with occasional dips down below 100 metres. The first officer and I agreed we would be lucky to get in on the first attempt. However that would be later. We needed to get through Rome first.

But our later-in-the-day challenge quickly turned into a real problem when we arrived onboard the Boeing 737-800. As I reached the flightdeck, the first thing I checked was the Autoland/Low-Vis status of the aircraft. The entry in the Techlog hit me straight away. This aircraft had been downgraded to CAT 1. Due to a technical snag we were unable to use both autopilots for an ILS approach therefore we had no Autoland capability. This pegged our minima for an instrument approach at 200 feet above the threshold (decision height) and at least 550 metres visibility. I turned to the FO,

"I will call Ops. Let's try and get another jet." Looking across the ramp there were another three 737s , so we should be in luck. It seemed to take an age before they answered the phone, but then a young woman's voice came on.

I explained the issue and asked her what they wanted us to do. I also said I could see three other aeroplanes down here in Sicily on the ramp, maybe we could swap? We were the first machine of the Dawn Patrol scheduled to fly. Naturally it would take a few minutes longer to change onto another aircraft, but I could see this would resolve the Low-Vis approach issue for later in the day. There was a pause while she spoke to somebody else, then she came back on,

"Just continue to Rome and we will see what the Turin weather is like when you get back to Catania."

I was stunned. These days we must never forget that the phone lines are recorded. I was very careful as I queried what I thought she had just said.

"Erm... you know the weather in Turin is forecast below Cat One all day right? I mean there's no chance we can get in there without a Cat Three aircraft, so why don't we swap now while we can?"

"No Captain. Thanks for your call. We will monitor the situation" and that was it.

The FO could tell from my face as I put the phone away, it was not good news. I confirmed it when he asked,

"What did they say?" To which I slowly replied,

"Err... she said... they will... MONITOR the situation!" which I really could not explain.

Instead I said to the dispatcher (whom we had already cued to expect an aircraft swap),

"Let's get the Rome passengers onboard, we're going!"

As we carried out the preflight I kept shaking my head in wonder. Maybe I should have been more forceful? Perhaps I should have asked to speak to the Ops Supervisor? But then, by her authoritative tone, young though she was, I think SHE was the Supervisor!

Judging by the bad language coming from the right-hand side of the cockpit, the FO was even more dismissive of our Operations Team. Surely they will arrange an aircraft swap ready for us when we get back to Catania I thought. Yes, that's it, they don't want us to think it was our idea. We will swap for sectors three and four. *Even though doing a swap to another ship halfway through your day is messier, but there you go. We can do this.*

The Rome Fiumicino sectors went well and we even arrived back in Sicily a few minutes early. We shut down on the ramp and disembarked the passengers. We were the only aircraft there. We checked the weather in Turin – still no change, the place was 'socked in'.

I called Ops again. This time it was a guy who picked up. I explained that we would not be able to land in Turin due to the fog and our Cat 1 aeroplane. I asked if there was any way to get an aircraft change.

"Standby..." He said, then a short pause, "no, there are no aircraft due into Catania in the next couple of hours, so go to Turin anyway" Now I really was incredulous. *Had I entered some sort of parallel universe where airlines don't function normally? I mean... what the f....?*

So, a couple of yoga breaths later I asked politely,

"When, we DIVERT from Turin, where would you like us to take the aircraft please? Because we will not get in. We cannot legally commence an instrument approach due to the weather being below our Cat One minima."

"Err... Standby!" then he finally said, "okay, if you can't land in Turin, then Pisa is the best place for us. Land in Pisa and we will bus the passengers from there. It's only four hours by road."

I shook my head, but agreed that we would do that. By now the FO was laughing and even the Senior Cabin Crew member thought it was so stupid it was funny. When the fueller asked us how much gas we needed. We looked at the flightplan and added another two tonnes to the ramp fuel, because today had become so ridiculous.

"We can forget the minimum fuel policy with this kind of decision making going on in HQ!" and the FO concurred.

"It's like the blind leading the partially sighted!" said the FO.

"Yep. Lions led by donkeys mate!"

Now just to cover ourselves, I called the Duty Pilot just before we pushed back. I wanted him to know exactly what we were doing and why. He could not offer a decent explanation either. That was a shame because I was rather hoping he would be able to share 'The Big Picture'. Maybe there was something we had missed?

We were now running a little late on our departure time, so I finished talking to the DP and closed the phone. The latest METAR from Turin was now down to 50 metres RVR (Runway Visual Range) in fog. Of course there was no point in informing the passengers they would be seeing the leaning tower of Pisa shortly. How would I possibly explain that? *Why did they not let us swap jets this morning when we had the chance?*

We were still laughing as we headed up the Mediterranean past Rome. We informed Pisa we would probably be diverting to them shortly. *There was little point in beating about the bush.* They expressed surprise, but confirmed their weather was wide open. We now reached our top of descent point for Turin... The latest RVRs showed a marginal improvement to 250 metres in all three segments, Touchdown, Midpoint and Stopend. We still needed 550 metres minimum to start an approach. We had no chance with our Cat 1 ship. Damn!

We commenced our descent, almost immediately calling for a diversion back to Pisa now just over our right shoulder by about a hundred miles.

"Erm... Standby! COORDINATION IN PROGRESS!" came back over the radio and then we heard *"FRANCO!"* on 121.50 of course.

"We can turn that idiot down," I said, selecting reduced volume on the emergency frequency on our second VHF radio. The last thing we needed now was any distraction. Handing over control and communications to my colleague, I spoke to the cabin crew on the intercom confirming our diversion to Pisa. They were not surprised. I asked them to listen carefully to my English PA and then to make an exact repeat in Italian for the passengers.

"Due to weather, fog, beyond our control, we are unable to land in Turin. We will be diverting to Pisa and there will be more information on the ground. Thanks for your patience and cooperation, we are disappointed also."

I could sense their regret through the flightdeck door. The cabin was then prepared for landing in Pisa. The ATC coordination came through. We turned lazily to the right with the autopilot as we continued descending and confirmed our set up for the landing in Pisa. Radar vectors and a nice handover to Pisa Approach, then we were onto finals for 04R. The landing was uneventful and we taxied onto stand.

After all was shut down and the forward door opened with the airstairs down, I went back into the forward galley and picked up the PA microphone. It is not something I like to do, but is a necessary part of our job. When the you-know-what hits the fan, the customers have a right to hear directly from the guy in charge. I said in very clear slow English that we were sorry about the fog in Turin and we would now find out what was going to happen.

I deliberately avoided making any promises regarding aircraft changes etc. *We could get an aircraft swap here and make our arrival into Turin if the RVR holds up.* The trouble was that now it was mid-afternoon, just after the highest temperature of the day. The Vis would deteriorate now as the temperature dropped. The OAT and dewpoint were the same anyway, there was nil wind and was no chance of any weather front coming through for the next 24 hours. We were up that creek again!

Then the airline station manager for Pisa came up the airstairs interrupting my thoughts.

"No problems Captain. We're all ready for them." He said, "get the passengers off into the buses here. They will go through the terminal, collect their bags and coaches have been arranged for Turin!"

Wow! That was easy. I again addressed the passengers from the front, explained in slow, clear English what would happen and apologised again on behalf of the airline. The Senior Cabin Crew repeated it all word for word, including my apologies. Although I think she may have used "Spiacente" instead of my heartfelt "Mi-scusi, Per favore."

The passengers were very stoic as they left the plane. I did not envy them coaching it all the way to Turin. The evening traffic would be building up by the time they got on the coast road. It would be five hours plus.

We all heaved a sigh of relief as the last passengers went down the airstairs. What next? I got on the phone to Ops again.

"The plan is you will change aircraft now and attempt an approach at Turin." The controller said, "we can see the weather's on the limits, but you might get in. If you do, then take the returning pax back to Catania. If you can't get in, then divert empty back to Catania"

I repeated it all back to her,

"What about our original pax, do you not want us to take them to Turin" but she was firm on that one.

"No, they've had enough for one day. We can't risk diverting them back to Pisa a second time. The adverse PR would be horrendous. They will go on the coaches now," and that was that.

We swapped onto a nice new ship, with full CAT 3A capability and loaded plenty of fuel onboard. Bearing in mind we might be trucking back down the length of Italy empty to Sicily again, then blasted off for Turin. The FO and I did a thorough job of preparing for the CAT 3A Autoland approach and briefed the cabin crew that they would not see anything until we landed. This proved exactly right.

As we reached our Radio Decision Height of 50 feet, only then did we see the runway lights for landing. It was right on the limits and could have gone either way. As we turned off the runway, I felt sadness for our passengers from Catania, who would only now be leaving Pisa for the long coach ride to Turin.

We taxied slowly in the fog and found the docking gate. We didn't need to refuel, so we started boarding the passengers for Catania as soon as the cabin was ready. We pushed back from the gate just under two hours behind schedule. Not bad considering everything. I was still mystified as to why they did not take my suggestion of 6 am this morning to swap planes.

I shrugged and said to the FO, "We just don't have the big picture."

✈

BACKSTORY: Those poor passengers never knew they had been the subject of incompetence. If the Ops Supervisor had listened to the suggestion of a veteran Line Pilot it might have been a different outcome. More often it seems that we, the Line Crews are disregarded with respect to our ideas. I wrote the whole story on the Journey Log (Captain's Voyage report) for the Flight Operations Management to read and received no response.

When I was a Flight Ops Manager at easyJet back in the early double ohs, I tried to incorporate the Captain's comments and lessons learned back into the system. I recall one of the best comments I ever read came from Captain Phil Dibbs. The former military (RAF) officer did not mince his words. He ended a brief synopsis of poor operational support with the words,

"It is impossible to 'Soar Like an Eagle' when you are operating from a TURKEY FARM!"

✈

37 INTEGRITY

If you had to choose the primary character trait for a professional aviator, you might think it would be flying ability. Or maybe physical fitness, or communication skills, but I think you would be wrong. Naturally these are all important personal attributes for a highly technical occupation. Without these capabilities the pilot would not be considered very good at their job. In my opinion they are all secondary to the most important one of all – integrity.

Without this, the pilot could misrepresent themselves in many different ways to disguise their lack of flying ability, fitness, situation awareness etc. They must be true to themselves and to others right from the beginning. In my own case, I was lucky. When I started my flying training, it was with the UK military. In that system, there is nowhere to hide. You are either good enough, or they remove you from the programme. It is as simple as that. In the early days of Royal Naval College, it was known as being 'Woofted!' – which bizarrely was an acronym for WFT (Withdrawn From Training).

Later on during the flying training courses, it was candidly known as 'The Chop'. I recall at RAF Linton-on-Ouse in 1984, one of our flight instructors standing in the bar of the Officer's Mess predicting, "The axe is going to go round like a Catherine Wheel next week!" He was not wrong; the following ten days were not pretty. Several of my course mates on the Jet Provost got chopped, I kept my head down and on my shoulders. I distinctly recall one of the instructors doing his best to fail me during that period while I was 'flying' the Jet Provost simulator. This wretched device had no motion, so it was a complete b***h to fly on instruments, but that's what we were required to do.

On this sortie I had lost my primary instruments due to failures. I was performing a limited panel instrument approach, using very little now that I come to think of it. I had already transmitted my Pan-Pan call and was coming down the approach completely blind. Except I had the turn and slip indicator, the altimeter, E2B magnetic compass and something else. Probably the clock! Needless to say, the selective radial instrument scan was a challenge. There was a crosswind and it was not easy, but I was on the glidepath and on centreline, being talked down by an ATC Warrant Officer.

All of a sudden at less than 500 feet above ground, the Engine Fire warning illuminated! We did only have the one engine. I merely transmitted the Mayday call and continued the approach. In fact, I think I even shrugged my shoulders, this was clearly not my day. It was too low and too late for ejection, plus there would be no Go-around from this one. Committed to land, I was only visual with the runway at minimums. It wouldn't have mattered if I had still been in cloud. This is one of the only situations when a pilot would continue below minima while flying on instruments.

As soon as the maingear touched down, I shut off the fuel, hit the engine fire extinguisher and the brakes simultaneously. Halting on the imaginary runway, parking brake set and switches all off, I started to unstrap to evacuate. I was very sweaty in my flying suit but pleased to be 'down'. I recall the Warrant Officer addressing my QFI, saying words to the effect.

"Bloody Hell Sir! That was a tough one! You put him through the mill there", and I smiled to myself. He was right, this had been no ordinary simulator flight. To which my young QFI blustered,

"Yes well. He's going through for single-seat fast jet, so they've GOT to be good!"

✈

I thanked my lucky stars that I had made a point of learning to fly on limited panel competently. It was Flight Lieutenant Bob Jones (a Master Green* QFI) who got me through my instrument training at RAF Linton. I am forever in his debt. When in the military not only is the bar set very high on every course, but there is continuous assessment while you are being trained.

Much emphasis is placed upon the verification of achievement. For example, the completion of military flying logbooks by each pilot. Perhaps it has changed now and maybe they have gone computerised. I am talking nearly 40 years ago of course. In the 1980s all personal flying logbook entries were completed first in pencil. The book was then submitted at the end of each month to the flight commander for assessment and when he was happy, we were allowed to enter all of our valuable flying hours in ink. Naturally the writing had to be very neat and perfectly legible. This was then signed off by our flight commander and stamped up with his ink stamp. We were indoctrinated with the thinking that our military flying logbooks were a valuable, legal document. They became our treasured possessions. The very thought of writing anything in our logbooks other than the sacred truth was anathema.

This schooling in truth above all was repeated later in our careers. Leaving the military environment we entered civilian aviation. We studied on courses to become trainers of other commercial pilots. It was made clear to us then that we must be honest in our appraisals and feedback to trainees so that they knew the areas in which they needed to improve. Later still when we became examiners on behalf of the civil aviation authority, it was incumbent upon us to judge performance fairly and truthfully against a set standard.

There have been times though in my career when I have faced difficult choices when working with colleagues who do not seem to share my view of the world. Maybe they came from different backgrounds. I am not trying to sound pious believe me. I am all for flexibility where necessary to give a hard-working trainee the benefit of the doubt. I am happier to pass people rather than fail, but a check should still be a check, right?

✈

Back in the early days of easyJet, I was scheduled for my annual line check and my check captain was the B737 fleet manager. He would be sitting on the flightdeck jumpseat for the two sectors out from Luton and back again. We briefed as a crew in the terminal and then went to the aircraft. Up until that time, I had assumed everything was normal. The first officer and I made all of our flight preparation diligently including the proper pre-flight and emergency briefing before departure. It was just after the passenger boarding had finished and we were about to close the doors. I half-turned in my seat to see that the fleet captain had occupied the observer's seat. He deliberately showed me his clipboard although concealing it from the FO.

It was the annual line check certificate with my name and rank at the top. He had ticked all the boxes as excellent, for both flights, including the Pass box and signed it at the bottom. I was shocked. Not only shocked, disappointed too. What was the point of us doing a line check now if he has already signed me off?! The Senior Cabin Crew leaned over his shoulder, confirmed the headcount and I told her the flight time and said she could close the doors.

Now it is not possible for you to tell your boss that he is being stupid so I merely smiled at him and nodded as if in appreciation. In reality I was seething. Up until that very moment I had thought he had integrity. Now he had lost all my respect. It is not a good way to start a normal flight, let alone what is supposed to be your annual check flight to assess your adherence to Standard Operating Procedures. In due course we completed the flights and I had passed my annual line check (obvs). However our relationship was never the same again. From that day I never took anything which he told me seriously. He had blown his credibility as far as I was concerned.

There have been a few other instances in my civilian aviation career when otherwise respected senior pilots have surprised me by their lack of ethics. Once when I was a TRE (Type Rating Examiner) on the Boeing 757 for a company based in Gatwick, I received a telephone call on my mobile from the Head of Training.

He said to me. "James you are in the Sim tomorrow doing an LPC with Captain _______. He's not very good and I would be obliged if you would fail him!" No kidding, those were his exact words. I could feel my blood start to boil, but I remained as calm as I could. Many thoughts went through my head. I replied with as icy a tone as I could manage.

"Mark, when I conduct an LPC it is a Licence Proficiency Check on behalf of the Authority. The check is done against the standards of the UK CAA and Captain ________ may pass or fail, but that's what happens ON THE CHECK. Now I'm going to put this phone down and forget this call ever happened, okay? Have a nice day."

That was it, the subject was never mentioned again. However, I now knew the man who was supposedly my boss had few moral scruples. I never trusted him again. Captain _______ actually passed his LPC the next day. He was not a brilliant pilot but he was competent and that is what matters.

In relating these incidents, I only indicate that there are a very small minority of people in senior positions whom I have met who lack integrity. The vast majority of professionals in commercial aviation I have known to be honest hard-working individuals, but there have been a few wrong 'uns.

BACKSTORY: I do not make any of this stuff up. I don't have to, it all happened in real life. There is much more of course, some of it I am unable to publish, but a lot of it I can de-identify and then go to print.

**Master Green – the top Instrument Rating in the RAF, it goes back to the 1950s.*

Line Check B757 - It is easy to fly from the jumpseat

38 A VERY EXPENSIVE DUCK

It was the weekend and I was just sitting down to lunch with the family at a local restaurant. I noticed they had duck on the menu. My mobile phone rang it was the chief pilot. He sounded irked.

"James did you not see the company SMS message about the emergency diversion from Liverpool?" I confessed that I had not, so he read it out to me. One of our Liverpool based aircraft had suffered a major engine failure immediately after takeoff and had diverted into Manchester. This occurred only minutes earlier.

I made my apologies to the family, asked them to get a cab home and ran out to my car. Fortunately, we lived only twenty minutes from Manchester so I would be on the scene quickly. As the recently appointed Base Captain for Liverpool, it was expected that I would attend to assist the crew.

I received more information as I was driving because the captain of the diverted flight called me. Apparently, they had just got airborne from runway 27 at Liverpool in poor weather, when all of a sudden 'Bang!' the right engine suffered catastrophic damage. Flying straight into thick cloud at 100 feet they cleaned the aircraft up and declared an emergency. This resulted in their landing on runway 05L at Manchester. What he did not mention to me then was they only flew for nine minutes...

✈

On arrival at Manchester Airport our handling agent met me and escorted me airside. The aircraft was remotely parked, so a short trip in a baggage truck was necessary to reach it. All the passengers had disembarked and the cabin crew were leaving to return to Liverpool. Only the captain and first officer remained with the aircraft.

As I got out of the baggage truck, I could see the damage to the righthand engine. My God what a mess! The N1 fan had partially destroyed itself and in doing so had taken the acoustic liner of the jet engine with it. It looked like part of a blade had broken off and then become trapped in the intake as it continued to rotate at high speed. The intake was shredded and other blades were severely damaged.

I went up the airstairs to find the pilots. I thought they might be still in shock from their harrowing experience only an hour earlier. The first officer actually appeared rather bullish I thought, but the captain was quiet and withdrawn. They related parts of the story, but it was all out of sequence. I told them we would get it down on paper back at the office. I asked them to secure the aircraft completely, made sure the Techlog was signed off correctly and urged them to get their stuff together. The engineers were on their way, so we headed for the Liverpool Base Office.

As they collected their personal belongings from the flightdeck I went downstairs and stared once more in awe at the wrecked CFM56 engine. What on earth had caused it? All those smashed and bent N1 fan blades. Surely it had been a bird strike. Then I saw it. A small brown feather on the inside lining of the air intake. *Just the one feather, from a duck perhaps?* I picked it off and turned it over in my fingers. So much damage, I looked again at the destroyed engine. *There must be plenty more inside from where this came from,* I thought to myself and absentmindedly dropped the feather which blew away across the concrete. I did not realise at the time, but I had just discarded the ONLY feather that would be found with the engine. I was kicking myself later – what an idiot!

On arrival at the office, I got them both to sit and write the Air Safety Report of the event together. We then emailed the scanned copy to head office. I recall at the time feeling sorry for them both. This is precisely the scenario we rehearse in the flight simulator every six months - Engine Failure On Take Off. In their case, it occurred just after takeoff, but still in a very dangerous position. The aircraft is at its slowest and heaviest. Not only that, but both engines are running at a very high (takeoff) power setting. A loss of thrust on one side is dramatic. I was pleased that they had succeeded in landing the aircraft safely at Manchester. It could have been a lot worse.

After they had departed I locked the office and drove home. The family had been unimpressed when I had dashed off, leaving them to have lunch without me. No matter how I tried to explain it, they did not really understand. I received a note from the chief pilot sometime before Monday, stating words to the effect that head office would take charge of the investigation. All I needed to do was to inform the crew that they would go to London by taxi at the first opportunity. At the time I assumed that it was because of the seriousness of the event. I did not suspect the real reason then.

The engineering team working on the aircraft had contacted me to say they had found nil bird evidence in the engine. First thing on Monday morning I contacted the bird control officer at Liverpool to ask for another inspection of the runway and surrounds to search for bird remains. This time I would accompany him. Early Monday morning I put on my high visibility waistcoat and met him airside on the ramp.

He coordinated with the tower by radio to have access to the runway and surrounds with his four-wheel drive pickup truck. While he drove slowly along the grass verges at the edge of the runway, I stood up high on the door sill holding the passenger door open to get a better view of the ground. Birdman of Liverpool kept saying he believed the bird strike culprit was 'wildfowl'.

This got me thinking, I wondered why he was so keen on wildfowl? Then I understood. If it was proven to be one of the species he should be controlling, he would have some difficult questions to answer. He had recorded flocks of wild duck flying through the airfield during the previous weekdays. On balance it must have been a heavy bird and a duck fitted the bill.

We must have spent the best part of an hour searching and covered a lot of ground. The birdman said that he believed the duck remains must have been removed by a fox. That explanation was logical and eventually we called off the search. I still felt embarrassed about that damned feather.

On arrival back at the terminal, the ramp safety officer was there. We had met before and I was not overkeen on his personality. 'Jobsworth' summed him up well. He informed the bird control officer that he would be issuing him a driving ticket because his passenger had not been strapped in while the vehicle had been moving. I tried to explain what we had been doing (which he knew anyway) but to no avail, the birdman received a penalty ticket and I shrugged my shoulders. I was profuse in my thanks to birdman and returned to the office for more bad news.

✈

This was a time before use of the Operational Flight Data Monitoring (OFDM) system had been implemented in the airline. Because there were no established protocols agreed between pilots and management, it was not possible to use the data formally in an investigation. That did not stop the airline's senior engineer calling me to ask how I thought our Liverpool pilots had handled the situation. I replied that I thought they had done well from what I had been told by them. I had taken their initial report at face value, thinking nothing was awry. He said he had listened to the Cockpit Voice Recorder (CVR).

He was alarmed with what he had heard. According to him, the engine had been running at high power destroying itself for a full six minutes after the initial bird strike. This in itself was not good news, but then he informed me that the crew landed only nine minutes later in Manchester not having completed any emergency checklists properly. The crew failed to mention the QRH* and he stated the words used from one pilot to another to indicate memory items were required to secure the damaged powerplant were,

"SHUT THE F****R DOWN!" Cross confirmation of the correct sequence as each engine control being moved was also absent. It was at this time I began to feel grateful it was not my investigation to run... The senior engineer went on to say that contrary to industry practice and company SOPs, the first officer had carried out the single-engined landing! The first knowledge the cabin crew had they had landed in Manchester was when they opened the flightdeck door! Clearly there had been a breakdown in good CRM in addition to a lack of operating competence.

✈

The result of the investigation with the crew was the pilots would return to the simulator for a couple of days retraining on Engine Failure On Take Off drills and CRM. The cabin crew were debriefed by the base cabin crew manager and reassured that the CRM standard of our airline's pilots is normally better than that.

I was informed by the training department that the attitudes of the captain and first officer were very different during the simulator training.

Apparently the captain accepted the analysis of the failings of the flightdeck crew, but the FO remained convinced they had done the right thing. Even though the captain tried to demonstrate his competence during the sim training, it did not go well. After the first session he telephoned the chief pilot and offered his resignation which was accepted. The first officer continued the next simulator session with a different captain and passed. He returned to flying the line.

It was with much sadness that I accepted the company property from the captain in the office the following week. He came in to say goodbye and return his company ID card, phone etc. I expressed my sincere regret that things had not turned out better for him. He was a real gentleman and did not once blame his pushy FO for what had happened. Privately the rest of us had our opinions on that score. The over-confident first officer who did not follow the SOPs, had not only trashed an engine, he had done the same with his captain's career.

The total cost of repairs to the engine was just under £1 million. In the Liverpool Base office we all agreed. It was a big bill for a duck.

✈

*QRH – Quick Reference Handbook. Contains all the emergency checklists.

Backstory: This all happened just as described. It was in my first few weeks as the Liverpool Base Captain. I was so relieved I was not involved with the investigation. One thing became clear to me. It is vital to establish the facts when analysing any serious incident and the only way to do that is to gather all the evidence. Because the aircraft had landed safely and nobody was injured, I jumped to the conclusion the pilots had done a good job. I was wrong.

✈

39 GO-AROUND, GO-AROUND!

Flying B737-800NG over the Greek Islands

On average a short-haul pilot will go-around once per annum and a long-haul pilot may do so once every two or three years. Of 1000 approaches to land with commercial flights, only two will result in go-arounds. Not only are they rare, but rarer still would be a day when you had to perform a go-around twice! This then is the story of that rarest of days in an aviation career.

✈

There used to be some different names for a go-around. In the military it was called an 'overshoot' and in the civilian world 'aborted landing'. Regardless, the whole aviation world standardised the description of 'go-around' long ago. It describes when an aircraft on final approach to land, breaks off the approach and climbs away again to make another approach to land. The go-around is carried out because it is safer than continuing to attempt the landing. There are several reasons.

The pilots may be unable to 'see' the runway – i.e. They reached their minimum altitude on an instrument approach and are still in cloud. Or maybe the runway is obstructed by a vehicle or other aircraft. The ATC controller may instruct the aircraft to go-around. Another possibility is that the aircraft is in an unstable state and not in the correct configuration for landing.

This last reason has been the cause of many avoidable accidents through the years. There are several ways to classify 'unstable' on final approach, but basically think in terms of an aircraft being too high, too fast or both. This would lead to an unsafe landing. I remember when there was some stigma attached to 'going-around'.

Some old dinosaurs stated with pride, 'never had to do a go-around in the whole of my flying career'. This implies that every one of their approaches to land had been perfect. Clearly that is nonsense when you are talking about flying aeroplanes for decades. However, it is indicative of the fact that traditionally performing a go-around was an admission of failure. It left me wondering how many bad landings these guys had made which could have resulted in the plane going off the end of the runway.

In my own case, I count myself lucky to have lived through the era when go-arounds became not only socially acceptable in aviation but were actually the safest form of manoeuvre to rescue an approach which went wrong. Personally, I have done quite a few myself, but it is still a rare event.

✈

It was a pleasure to see first officer Andreas' smiling face in the crewroom that morning, we had operated together previously. I knew him to be a competent aviator with a good sense of humour. Ours was the first flight departing at 0600 local time out of Athens, Greece. The Dawn Patrol. We were flying a Boeing 737-800. As usual the weather was lovely, with plenty of blue sky and light winds. A typical high-pressure stable weather system over the Greek islands. Our schedule was four sectors from Athens. First we would fly to the island of Mykonos and back, then Rhodos and back.

When we looked at the weather together in detail, we noted that Mykonos had some partial cloud cover in the forecast, but the rest of the islands were clear. We did our brief with the cabin crew, then headed out to the jet. It was such a beautiful morning in Athens, slightly crisp, but with a clear sky. One of the things I enjoyed most about flying short haul from Athens was the incredible views of the islands in the Aegean Sea. I knew today was going to be a beautiful day.

As we prepped the aircraft for the first sector, I checked the latest report for Mykonos again. It was showing 'scattered cloud at 600 feet' with very light winds. No problem. They would be using runway 34 for landing and we might even get a visual approach if traffic permits. Little did I know what awaited our arrival a short time later.

In our profession we are always pleased to highlight our enviable safety record. I know it does not always seem that way, because the aviation industry is so high-profile, but statistically it is the safest form of travel. This is to the extent that after a couple of decades flying airliners you become aware of the large safety margins which exist operating in and out of large international airports. It is relaxing for example to fly into London Heathrow, even at the busiest of times.

Why? Because everything is so well organised. The air traffic service is one of the best in the world, everything works properly, navigation aids, lighting systems, ground movement radar, docking installations, you name it, it works! “Straight down the ILS, RW27R, cleared to land, follow the greens, parking stand 214.”

There are some other destinations which are not so relaxing. They are well known among professional aviators as being places where you can immediately name threats and are already thinking of ways to mitigate them. They are ‘special’. Gibraltar, Calvi, Salzburg, Funchal, Madeira and the Canarsie Approach for runway 13L at JFK come to mind. There are plenty of others. These places stand out in our memories because this is where the safety margins are reduced.

It is still ‘safe’ to operate there, but you must pay attention to ensure correct procedures are followed including all the special limitations. The runways are shorter and/or narrow. There is high ground nearby, the weather has idiosyncrasies, there are no precision approach aids to use, the parking ramp is tiny, the potential threats are numerous. This makes such airports more challenging but also more interesting.

Mykonos island airport is one of them. It always receives our full attention. The runway is short, it is also narrow, there is high terrain very close. There are no precision approach aids. There is only the VOR radio beacon for runway 34 with high decision minima for the approach. For runway 16 there is only a visual approach with nil PAPIs. It is a ‘Captains only’ airport for jet operators, meaning that company restrictions do not allow the first officer to operate there as Pilot Flying. There is no radar and the ATC controllers have their own ways of operating sometimes which can be a surprise if you were expecting a Heathrow type of service! Oh, and when the runway is wet, it is slippery due to the amount of rubber left on the surface by aircraft braking heavily.

With all the foregoing in mind we were preparing to fly there on this fine day. I spent time setting up the flightdeck for departure, while Andreas completed the external checks and the fuelling. He then joined me and we completed the briefing for the flight. As the flight from Athens is less than 20 minutes, I also briefed the full instrument arrival procedure for runway 34 at destination. From experience the more time you spend on the ground briefing the known elements of a short flight the better.

Mykonos island airport is under there somewhere...

We were in the descent towards 'Party Temple' as the island is known among young Athenians, when I could see that the cloud layer was much more than 600 feet scattered. Over the open sea there was little cloud, but over the island it looked more like broken or even overcast at the southern end. We were both amused at the optimistic Met report, but we continued to prepare for our approach. Unlike previous generations of airliners, the modern glass cockpit typified by the Boeing 737NG means that we have a much better mental air picture in front of us.

No longer do we have to rely solely upon a VOR radio needle to show us our relative bearing to the beacon and a DME* readout to show distance from it. Now in addition to Inertial Navigation Systems, we have GPS to show where we are in the sky, plus of course all the modern terrain warning systems. On handover to Mykonos tower, we asked them again for the latest weather and they confirmed '600 feet scattered', but without any mention of higher altitude cloud amounts.

As the first aircraft to arrive this morning, there were no previous reports from other flights to help us – we were on our own. The only element of the METAR which we knew would be reliable was that of the surface wind. They could read that off the instrument in front of them in the control tower. Mercifully it was less than five knots - light and variable.

We spoke briefly to the cabin crew, explained about the weather and advised them that a go-around could be a possibility. They would not be so surprised if we had to climb away from the first approach and try again. It was at around 8000 feet in the descent that I took a photo with my iPad for posterity saying,

"600 feet Scattered what a joke!" We were cleared for the 'VOR approach runway 34, call again on finals'.

I slowed the 737 right down, we were not going to rush this one. I called for 'Flaps 1' early and then 'Flaps 5', shortly after that I called for 'Gear Down Flaps 15' as we flew over the top of the island outbound from the beacon. The autopilot was doing a good job, allowing us to monitor our progress as I dialled in the next altitude 2000 feet, the minimum at the end of the leg out from the beacon and in that right hand turn on to final. We levelled off in the turn, over the sea at 2000 feet. We were IMC in solid cloud. *If this continues all the way to minimums...?* there was every possibility of a go-around coming up.

"Approaching Descent Point!" came nice and clear from Andreas. It was so reassuring to know I had somebody in the right-hand seat who I could rely on. *He's got our back, good man.* I replied as per the SOP.

"LNAV, VNAV PATH, Speed Intervent" pointing at the MCP windows as I did so. Then I selected the minima of 1200 feet in the MCP window of the autopilot and shortly afterwards called for 'Flaps 25', we were entering our final descent. Now it was quiet, very quiet in the cockpit as we concentrated.

No noise from the cabin, nothing from ATC, just the two of us on the intercom. The engine noise had decreased. We were still in solid clag. A blanket of grey outside the windows. Andreas confirmed the altitudes versus distances for me, we continued descending, I now had both hands and feet covering the controls ready to take over manually and land the machine if we got visual. He called,

"Approaching Minimums!" we were only a hundred feet above now, still in cloud, then "MINIMUMS!" I immediately responded,

"GO-AROUND FLAP 15!" and simultaneously pressed the Go-around TOGA switch with my right index finger very hard! The thrust increased as the levers advanced and the autopilot raised the nose following the magenta flight director. My eyes were locked on to the attitude of the aircraft on the Primary Flight Display. I could hear the 'shhlick' and 'clack' of the flap lever as my colleague did his stuff and then as I could see we were climbing I called "GEAR UP!" and the reply came back,

"POSITIVE CLIMB!" as he pulled the landing gear handle out and into the UP position. We broke cloud into bright sunshine at 2500 feet and then reached 3500 feet our missed approach altitude and levelled off. We kept Flaps 5 extended, because the procedure required us to 'stay below 185 knots and turn at five miles' from the beacon.

This we did and it was in that left turn back towards the overhead we saw the full extent of the cloud to the south, it was overcast all the way to final approach runway 34. At this end (to the north) it was just clear, a visual approach to runway 16 would be possible. It really was 600 feet scattered up at the north end of the island, but that would change rapidly.

We had a quick discussion, checked the fuel and both agreed a visual approach to land on RW16 was our choice. We checked with the tower and they were happy to oblige. I gave Andreas control of the aircraft while I made a reassuring PA announcement to the cabin. This had the benefit of informing the cabin crew of our plan at the same time – we would try to land on runway 16. I took control again, looked up the approach plate for the circling approach while Andreas typed the procedure into the Flight Management Computers. I looked across the cockpit and was pleased to see his fingers were a blur of movement. *This kid knows his stuff,* went through my mind as I watched him. Once it was done, we briefed quickly for the other end of the runway and informed ATC we were ready.

As we were at Flaps 5 already, it was not much more to Gear Down Flaps 15, which is our best manoeuvring configuration for a circling approach. We started off in cloud, but as we turned onto the downwind leg out to the west the cloud broke up nicely. Heading north now, I kept checking with Andreas and he was able to confirm we were maintaining visual contact with runway 16 threshold. We maintained our circling altitude of 1700 feet.

Now as the LNAV traffic pattern took us closer, turning towards the island, I called for 'Flaps 25' and shortly afterwards I disconnected the autopilot calling 'Flaps 40' at the same time. Now I was concentrating on flying the best curving approach to the runway. Using the smallest power changes possible and gently on the controls, I was constantly reassessing our height versus range. We completed the landing checks and received landing clearance.

"Five hundred STABLE!" came from my right-hand man and we continued. He called out the wind. It was very light from the east. Now we were on finals and it appeared we were low. But I kept telling myself that was 'normal' as the land slopes down towards the runway.

Suddenly we were over the threshold and the auto-callouts came out loud and clear, '50, 30, 20' as I flared slightly and closed the thrust levers the maingear touched – BUMP! Immediately the 'Max-Auto' braking took over and I selected full reverse thrust. The roar from the CFM56s was great to hear as we slowed very rapidly on the runway centreline. We were down!

We really did feel like we had won something by the time we had shutdown and the passengers were disembarking. Nobody seemed perturbed by our go-around and second approach to land. In those days, it was self-manoeuvring on the apron there were no pushbacks at Mykonos and we were facing north. Looking straight ahead therefore we had a great view of the cloud rolling in which obscured runway 16 threshold from the air.

By this time the tower was sensibly broadcasting 'broken 500' and 'overcast 1500'. To keep our situation awareness high, we kept the tower frequency on a speaker in the flightdeck and it was on that we heard the Aegean A320 going around from an approach to runway 34. The cloud was impenetrable above us as we witnessed the roar of his engines climbing away to the north. When asked what his intentions were, the captain had no hesitation, 'we are returning to Athens.' we laughed, that was a first for Aegean.

Then a large private jet tried to land, twice. After some minutes airborne holding, we heard them eventually give up and confirm they were diverting to Paros - another island in the Cyclades, not far to the south. Finally we had our passengers onboard and were ready to depart for Athens. Ours was the only aircraft in Mykonos and it felt eerie as we backtracked runway 34 for departure. The cloud still persisted and there was no way any other flights would land while it was there. Not a problem for us taking-off though and soon we were hurtling along the runway approaching flying speed.

'BOOF!' we burst out of the solid cloud layer into a bright blue sky above. What a great feeling that was climbing fast towards Athens. On a direct track of course, there was no other traffic flying. We said our goodbyes and called Athens approach. Again the flight time was very short and in no time at all we were disembarking on the ramp and getting ready for Rhodos.

We noted only one NOTAM of significance, the Rapid Exit Taxiways, Bravo and Charlie from runway 24 were closed for resurfacing. No drama, we would taxi further along the runway after landing and vacate at the Delta RET. The weather in Rhodos Diagoras** airport was clear, CAVOK, so we had no worries now. In fact, the mood among the crew was euphoric, we felt like we had got away with a surprise 'win' at Mykonos. The next two sectors were Andreas's so I was in the Pilot Monitoring role now.

We were in the cruise towards Rhodos and I handed him the weather with a smile,

"All good mate, it's CAVOK, runway 24 with light and variable winds, it's a doddle!" Of course, I spoke too soon, but we had just experienced our one go-around of the year, so life was good.

We could see the island from nearly 100 miles away as we commenced our descent. Cautious as always, we planned the LNAV track to skirt around the border with Turkey only a few miles to the east. Soon after calling Rhodos radar, we were requesting a visual approach, it was such a beautiful day.

There was only one other aircraft in the sky, 'NorTrans' (a Norwegian Airlines B737) ahead and below us. They were now on finals for runway 24 and we confirmed visual with them for the controller,

"Clear to self-position visually behind the traffic on finals, call the Tower 118.20"

So now Andreas was calling for Flap 5 and we were slowing right down to come on to finals at six miles, plenty of room behind the landing Norwegian. They were just touching down as we passed the five-mile point, so we had plenty of room. The female voice from the tower cleared us to continue our approach. Now we were at four miles and they were on the runway, but they were slowing down too quickly, we could see that.

"Oh no! They're going for the rapid exit for the ramp, it's closed Damn!" quick thinking Andreas called,

"Flaps 40! Reset the Vee Ref!" and I selected this for him and we reset final approach speed to give us seven knots slower, but would it be enough? The Tower controller also saw what was happening and started transmitting to the Norwegian,

"NORTRANS 737 EXPEDITE! Traffic on finals!" but the dull crew on the Norwegian just thought she meant clear the runway and seemed to slow down more. Even we could see the orange and white cones blocking off the exit from the runway –

"Are they BLIND! Get ready Andreas, I think this is going to be another go-around mate…"

We were now inside 3 miles from the threshold and the Norwegian was only just accelerating again from nearly a full stop, to the chorus of,

"EXPEDITE EXPEDITE! THAT EXIT IS CLOSED!" then the words I had been expecting, "______ (our callsign) GO-AROUND! I SAY AGAIN GO-AROUND!" she was really shouting now and we responded immediately, she had left it late enough.

"Go-around! Flaps 15" came from Andreas as he was also swearing about idiot Norwegians, I laughed as I moved the flap lever, simultaneously transmitting,

"______ GOING-AROUND! YEAH - THANKS A LOT NORWEGIAN - *FOR NOT READING YOUR NOTAMS DUH!*" then I called on the intercom,

"Positive Rate!" and he replied, "Gear Up!" I looked across the cockpit and he was well in control of the situation.

"We're getting pretty good at this now aren't we hey?" which prompted him to laugh.

We taxied past the errant red-nosed 'clown-jet'

Back to the approach controller we agreed a,

"Visual circuit, short pattern downwind, you are NUMBER ONE to finals!" Even he could not believe it. I made a short reassuring PA to the cabin,

"Ladies and gentlemen, the runway was blocked. So now we make another approach to land. Enjoy the pretty views. There will be no extra charge for the sightseeing!" which amused Andreas again and he was giggling. So I said,

"Well! It IS a LOW-COST airline so they (the passengers) might be worried..."

Our second approach and landing were uneventful. We taxied past the errant red-nosed 'clown-jet' on the ramp. I was proud that we resisted the temptation to make rude gestures from the flightdeck windows. We were too focused on getting the turnround done efficiently. No surprises, we beat them on that score also. As we called for start-up, Norwegian was still boarding passengers.

*DME – Distance Measuring Equipment – the part of the VOR/DME beacon which gives nautical miles readout from its location.

**Diagoras – he was a famous ancient Greek boxer from the 5th century.

Backstory: Andreas left the airline not so long afterwards. We are still in touch; he works for Austrian Airlines these days.

The B737-800NG a very good cockpit display for go-arounds

40 AND THE BAND PLAYED ON

VIP B737 crew – everybody's got a heart (& a visa for India)

We had commenced our descent into Yerevan. From listening to the latest ATIS broadcast we knew it was snowing at the airport. This was not good news. In the flightdeck, I was concerned. We really needed this techstop to go well. The airport in the Capital city of Armenia was halfway to Jaipur in India. Jaipur was our final destination and we were flying an old Boeing 737 to the limits of its range on both sectors. Not only that, but this was a VIP flight with 50 such customers in the back. If it went wrong, there was nowhere to hide – our VIP passengers would want someone to blame. That would be me.

✈

Thinking back, it had been a struggle all the way with this one. From the original booking of the charter, through to getting the crew together with all their visas ready for India. Talk about leaving it late! I was still in central London the day before the flight getting my own visa issued at the Indian Consulate. By that time I had gone beyond stressing about it.

I was in the fatalistic phase, where I had adopted the attitude of 'if it happens, it happens'. Well, it DID happen! The precious visa was stamped into my passport and I mistakenly took that as a good omen. I really should not be such an optimist.

Getting the aircraft ferried into Luton in advance was not without dramas. There were some technical snags which the onboard engineer was still trying to fix even after the ferry flight had been completed.

Now the 60 seat VIP airliner was parked on the ramp outside Harrods Exec Jet Centre looking magnificent, but behind the scenes it was all happening! Problems with catering, problems with mechanicals, issues with permits and permissions, plus other flight-planning obstacles and finally the weather forecast looking dodgy.

At least the crew were all there 'booted and spurred' and I had managed to score my visa. Which was just as well because we needed three pilots to achieve this particular mission. The extended length of the duty times meant we were operating 'heavy-crew'. With three pilots we could take inflight rest in the VIP cabin. Not so the poor cabin crew, there were only four of them and they would be mostly on their feet all the way to India.

I had two experienced B737 captains flying with me who had flown many VIP flights before. We knew each other well and worked well as a team. This would be quite a mission. The principal had chartered the VIP airliner to take his family and friends halfway across the planet to his daughter's wedding weekend in India. Moreover the wedding would be attended by some very senior government officials. Our client was fabulously well connected. this would be a very special occasion and our flights were an important part of it. Meanwhile as I was rushing back across London to get up to Luton on a Friday afternoon, all I knew was that the aircraft was not yet ready.

Imagine my relief as I finally walked into the hotel at Luton to find my crew all sitting there with big smiles on their faces. The technical snags were cured, catering issues had been resolved with the help of the lovely people at Harrods Executive and the pilots assured me they had seen evidence of all the overflight permits. Now it was time for a relaxing evening with a late start planned for tomorrow. The flight was not due to depart until 6.30 pm, so there was no rush to get out of bed.

After our planned departure from Luton at 6.30pm, we would fly through the night, to land in Jaipur at 9.30am the next day. It was quite exciting being a part of the preparation for this little adventure. Leaving all the hassles aside, making a VIP trip like this happen and run smoothly was something we all got a kick out of. Much input from the crew is needed to make these flights successful, so it is satisfying when we solve all the issues. It is like a jigsaw puzzle being completed by a large team of people. So pleasing when it all comes together.

Now it was late afternoon on the day of our departure, we had finally arrived at the aircraft and were deep into pre-flight preparation. As always the cabin crew looked completely different in their uniforms and make-up. The pilots were looking immaculate too. The catering truck was loading, the potable water was being refilled and refuelling was in progress. Tonight we would need full tanks for sure.

In addition we had booked the pushback team to come at the right time and the baggage loaders were standing by. I noticed that the instruments for the band had arrived and were already in the cargo hold. As part of our charter flight we had agreed to carry the musical instruments for the wedding band whose members were flying out with BA from Heathrow. Naturally nobody had thought about any customs formalities...

As I walked into the executive lounge just in front of the nose of the jet, I could sense the visibility had reduced again. The temperature and dewpoint were getting closer all the time. The fog was coming. Even back in '97-98 when I worked at Luton for easyJet we had to battle with the fog. We used to say then, 'Who in their right minds would build an airport on a hill in cloud?' The elevation of Luton is 500 feet above sea level. It does not sound much, but it is significant. We needed at least 400 metres visibility to permit our takeoff as we did not have Low Visibility Operations approval.

*"F****** Google Earth? Seriously!" he was not impressed I could tell*

For us pilots the tension was rising now with each reduction in the reported RVR (Runway Visual Range), it was down to 700 metres, we could not afford much further deterioration. I had already spoken to our head of operations on the phone to check if there was any way round the visibility restriction – the answer was 'no'. Imagine having to cancel the flight because of departure airport weather? I shuddered at the thought of facing 50 VIPs with the bad news. Inside the building however, all was perfect. *Harrods interior crew do their job well* I thought looking around the elegant departure lounge. The first of our VIPs had arrived and were settling in, we still had nearly two hours to go.

Stood there in my uniform chatting with the customers, I felt like a bit of a fraud really. The dodgy medicine man selling sugar water. 'Oh yes, the weather is fine, the aircraft is wonderful, we fly trips like this all the time...' In fact the weather in Yerevan was pretty lousy, the aircraft only just had the range to do the job and we rarely flew this jet outside Europe. Yes this was a big deal for us. What else could I say though? You have to have faith.

I knew we had a great team of people and I would not let them down by failing in the PR role. I just prayed the visibility would hold up. The OAT would not stop falling now all evening. I knew from experience that once the fog established itself that would be the story for the whole night. There was very little wind and no chance of a frontal clearance for a few days.

We had already decided that if we were going to cancel, we would do so with all the passengers in the lounge. It was unfair to subject the cabin crew to that disastrous situation. I excused myself as my mobile phone rang and moved towards the windows to take the call.

It was operations asking how it was going.

"Not good I replied, we are literally at the mercy of the elements here and it looks like it will be touch and go." They said the RVR was 'now below 600 metres and Luton has gone into Low Visibility Procedures'. My heart sank as I heard this, we still had an hour plus to departure. Then the charter broker waved to me across the lounge, he had the principal with him.

"Keep monitoring it, let me know if it changes, I have to go."

I went over to where they were standing and was introduced to the principal. He looked happy and relaxed. Much more so than I felt. As I trotted out all the usual platitudes I wondered if he would be so cool if he knew all that I knew – probably not. It was about this time I was informed that Guy Ritchie the film director was one of our passengers. I imagined how it would be, telling the writer and director of 'Lock, Stock and Two Smoking Barrels' he's not flying tonight? Not a pleasant thought at all.

Now the lounge was really filling up and the ambiance was lovely. Pleasant music in the background, the excited chatter of wealthy people travelling on an adventure together. I checked with the crew, the aircraft was ready, but the visibility had dropped again, now it was 400 metres our absolute limit. Damn! It was less than 30 minutes to scheduled time of departure, but this was looking really bad now. I felt sick. It was time to talk to the charter broker and let him know the worst. Then we would have to face the customers together. I could already see the bus he would throw me under!

Finally I managed to get him alone and explained the situation. He was aghast and I felt the same. I emphasised that if we had departed even half an hour before this we would have been okay, but not now. We were screwed. The last of the passengers had now arrived and were mingling with the other glitterati - there's no other word for it, they all looked amazing of course. This was a significant section of London high society going to India for the weekend, 'like you do!'

I said to the broker,

"All we can do is pray. If the visibility doesn't increase, we are stuffed. Let's delay telling anybody just yet and let's see what happens."

Which was actually a good idea. Because a short time later, I got another call from the flightdeck, it was Doug.

"James, you're not going to believe this. But the vis has increased. It's now 500 metres!"

At the time I was very close to the principal who was now looking at his watch, he had realised we should be boarding the flight by now. I caught his eye and smiled,

"Shall we go?" I asked him casually as if it was really his decision. He nodded.

"Yes, let's!" he replied draining his glass. And with that we commenced boarding the aircraft. It took a while as the guests all had to pass through security to go airside. Meanwhile we tracked the RVR reports assiduously, it was holding, just.

...ours was technically not a 'Low-Vis' takeoff

I was pleased to be sitting on one of the flightdeck jumpseats for departure and revelled in the fact that the two gentlemen in the pilots' seats were top drawer commercial aviators. The airport was working in LVPs but ours was technically not a 'Low-Vis' takeoff. It was above 400 metres RVR. I listened intently on the spare headset as we taxied out for the runway and in the cabin I could hear the safety PA going on as normal. It was a tight squeeze as all four flightdeck seats were occupied, Mark our engineer was with us too.

The fog blurred past the windows as we accelerated down runway 26. There was a collective sigh of relief from the four of us as PM called 'Vee One, Rotate' and PF raised the nose into a climb. We had got away with it by the skin of our teeth. All of a sudden, we were talking to London Control and climbing on track towards the east. Nobody said anything. There was just the clack, clack, clack of the autopilot trim wheel as the climb speed increased. It was going to be long night.

Three and a half hours later and I was looking at the weather report for Yerevan in the darkened flightdeck. Now I was in the front seat, ready to shoot the approach and landing. I had completed the briefing which included the fact that the SSA (Sector Safe Altitude) was as high as 18,100 feet to the south inside 25 miles. Even in the direction we were coming from the SSA was over 11,000 feet above sea level. We were planning to land at an unfamiliar airport in mountainous terrain, in the middle of the night, mid-winter in the northern hemisphere and it was snowing. It was a critical time. Stefan in the co-pilot's seat said,

"Tell me again how you chose Yerevan, Zvartnots airport as our techstop?" to which I responded,

"Well, you see. I looked on Google Earth and ran the straight line from Luton to Jaipur. And it was sort of halfway really and I was in a bit of a rush."

"F****** Google Earth? Seriously!" he was not impressed I could tell. But it had seemed like a good idea at the time. I knew that if it was still snowing hard by the time we landed, we would have to deice. We might not even get in if the runway was not cleared or if we could not see to land. I shook my head at the thought of that. Not a good outcome. Better just to get on with this and hope for the best.

"Tell them we're ready for descent mate. Let's go and see Armenia." The next 30 minutes were tense as we had the engine anti-ice on for most of the time. Horizontal snow streaked past outside. Occasionally, we saw the build of ice on the outside of the flightdeck windows, so we checked the leading edges of the wings with the lights. It was there too.

We agreed that it was important to leave it as long as possible before we activated the Wing Anti-Ice. This blasts hot air from the engines along the leading edges of the wings and heats up the slats to melt the ice. If you do it too early though, with just small amounts of ice build-up it melts, and the water runs back to refreeze on the wing where you cannot melt it off.

That is undesirable to say the least. So we waited each time until we could not stand it any longer. The we would hit the WAI switch, watching from the flightdeck as huge chunks of ice broke off and blew away in the slipstream.

Now with our final radar vectors towards landing, the clouds were breaking up a little. Occasionally we saw lights from towns through holes in the cloud. The snow had stopped and we were on the ILS approach to land on runway 09. Our Decision Altitude was set for 3000 feet on the altimeters

– the elevation of the airport was 2800 feet above sea level. We could breathe again when we saw the approach lights from five miles away. We were so glad this approach would not be down to minimums. Light snow grains were falling as we taxied off the runway to a remote parking area. In the flightdeck we were all smiling, we had done it. We were halfway.

One of the great things about carrying a travelling engineer, or 'flying spanner' as they are known in the trade, is that they will do the external checks for you. Mark was a smart guy and had brought his big coat, gloves, hat and scarf. He looked a bit like the Michelin Man as he went down the steps to monitor the refuelling while we put the kettle on in the forward galley. Again the fuel tanks were filled to capacity and we were pleased that it was so cold outside. Just below freezing would give us a good uplift as colder fuel produces a better Specific Gravity reading; we needed every kilogramme.

Another advantage about Yerevan was the ground handling and airport operations. They were very efficient and were busy running snow clearance teams on the apron and out on the runway all the time we were there. Even Stefan was impressed. I tried to pretend that I knew it would be this good, but he just laughed at me,

"Yeah right! 'Course you did. Mister Google Earth!"

Soon enough we were ready to go and pleasingly we turned it round under the scheduled 40 minutes techstop time.

...crossed above some of the most inhospitable areas of the globe.

We took turns through the next sector in flying the aircraft. We were grateful that the flying conditions were so smooth. We departed south into Iranian airspace and just south of the Caspian Sea the sun came up and hit us squarely in the face. In the cabin it was still dark and the cabin crew had been able to rest a little prior to their five-star breakfast service they had planned. Another hour or two over Iran and then we crossed the southern part of Afghanistan in Pakistan. It was a relief to hear 'Namushkar!' from the Delhi Air Traffic Controller. It was not long now. Breakfast had been served and we had all enjoyed our early morning coffee watching our beautiful planet earth slide past under the nose.

The Jaipur weather was sweet. No problems at all. A little hazy of course (this is India right) but with only light winds which enabled one of my two expert aviators in the front to make a very smooth landing. As we shutdown on the ramp at the local time of 9.30am I ensured that I was in the forward galley with the cabin crew to assist with disembarkation.

Mark was straight down the steps as soon as the forward left door was opened. He would supervise the orderly offload of the baggage, but not the musical instruments. For the sake of expediency and safety, we had agreed to take those through with our personal baggage to the crew hotel where they could be collected later. The weather was already hot even at that time of the morning. We looked forward to a pleasant weekend off all together while our passengers went to the wedding. Jaipur is an amazing destination and there is so much to see in Rajasthan which is famous for its palaces. As the last passenger left the aircraft we started our cleaning routine.

The cabin crew liaised with the catering company to offload all dirty crockery, serving dishes, cutlery and glassware. The pilots set about helping with the cleaning where possible, we were good at that by now and Mark checked the aircraft inside and out. I was pleased to hear that he was well satisfied with the technical status of our ship of dreams. We were a very long way away from home if we needed maintenance.

Finally we were done and pretty much done in! We had been up since yesterday lunchtime, had fretted over the low-vis departure from Luton. Worried about the weather in Yerevan and then crossed above some of the most inhospitable areas of the globe. Our amazing cabin crew had virtually walked all the way to India, with a few breaks to put their feet up. They had prepared and served a superb feast of a dinner last night to 50 people in an aluminium tube, then done the same with breakfast this morning. As I looked around at their exhausted faces, I knew this had been a big one.

We had finished making as much preparation as we could for our return flight to London on Sunday, a full three days from now. It was time to head for the Hotac. We gathered all of our personal stuff together. I asked the crew to go on ahead to the terminal while I stayed with Mark, they should send the bus back for us, we would follow in a few minutes. There was bound to be some delay going through immigration of that I was certain. Welcome to India!

We closed up the aircraft totally and fitted all the security seals. When leaving the machine for an extended period like this, we always stuck our special seals across every door and hatch of the aeroplane. Then when we returned we would know instantly if anyone had entered the aircraft. Every seal is individually numbered and these days it is easiest to run round all the stickers with your mobile phone to photograph the location of each one. I liked to do it personally so that I was sure it had been completed fully.

The cabin crew looked completely different in their uniforms.

The bus ride to the terminal felt hot after being in our air-conditioned aluminium tube for so long. Then we were back into blessed aircon again as we entered the building. I was pleased to see that the crew had gone on ahead as we approached the immigration desks. They would be waiting outside for us, possibly on the hotel bus by now. My new visa was noted by the immigration officer, although I am not sure that he got my comment about the ink being still wet from it being issued only hours before it seemed. With our passports stamped, we were free to go to the baggage hall and Mark and I were just remarking about how hassle free it had all been. That was tempting fate of course.

As we stepped onto the top of the escalator which led down into the baggage hall we were greeted by the sight of the whole crew sat on suitcases or benches looking most dejected. There were the trolleys with our crew bags, but then at one side a further four trolleys laden with the musical instruments for the band. Stef and Doug were the only ones standing.

They looked very glum and it was clear they had been having a long discussion with the senior customs officer. He was dressed all in white like the other officers. His uniform dazzled in the sunshine streaming in from the windows high up in the hall. His body posture said it all.

Arms folded and legs slightly apart, he and the other minions had decided the musical instruments would be impounded. The customs laws in India are among the most restrictive in the world. Stefan turned to me to explain, he looked drained like all the rest of them. We had come so far and now THIS!

"The instruments for the band are banned!" was Stef's way of announcing we had a big problem here. I went to Mister White Suit and asked what the issue was. You could tell he was pleased. He was doing his job one hundred percent and he had just managed to prevent the illegal importation, nay smuggling even, of some very valuable instruments and which he had listed on his clipboard. I smiled at him. This was disappointing for us, but we don't show that. It was time for the posh British accent.

"Where is the office of the HEAD customs officer please?"

"I am head of customs here Sir" he responded with a confirmatory head wobble. But I knew that was not the case. India is obsessed with hierarchy and status.

The one 'in-charge' would be sitting down somewhere in an office, he would not be walking around working like this guy. I was not to be dissuaded. We had been up all night and conquered so many obstacles to get here.

"Where is the OFFICE of THE MANAGER of the whole customs? You know, the 'IN-CHARGE', I need to see HIM please."

He indicated the office at the back of the customs hall with a plaque on the door. The blinds were down inside the windows so you could not see in – ah yes, that was the place. I walked smartly over and knocked politely on the door,

"Come!" was the reply from inside, so I opened it and there was a very nice office with a large air-conditioned space between the door and his desk. The senior officer was alone. He stood up from his chair behind the desk and I smiled - very posh now,

"Good morning Sir! I am the Captain of the VIP flight which just landed" I went in and closed the door behind me, "my name is James and here is my card. I believe you are head of Jaipur Airport Customs?"

He welcomed me although taken aback by my positive attitude. I was smiling throughout. *I will enjoy this* I thought to myself.

He sat down behind his desk with my card in his hand and motioned for me to do the same on my side. I sat and he began to explain about smuggling and importation laws covering the illegal importing of valuable musical instruments. I took out my pen and a piece of paper and nodded still smiling, as if I did not care one jot about the impounding of the instruments. After all they were not my instruments.

"Yes, thanks for that. I understand completely. But you see all I need to know, is how to spell your name? Forgive me my friend, but I am not good with all the Indian name spellings. Ahh! Here it is!" I pointed at the wooden and brass name plaque on his desk with his full title and copied it down carefully.

He was frowning now and I could see he was curious. So I explained speaking slowly, (always use the name...)

"You see Mister __________, our private jet brought 50 VIPs for a wedding in Jaipur this weekend. Your Prime Minister is one of the guests at the wedding along with other people from your government - my principal is VERY WELL CONNECTED. We brought the musical instruments with us for the band who will play at the wedding, but when they cannot play, my principal and your Prime Minister will be asking me where are the instruments? And...", I smiled a big smile as I stood up, "I will give him your name! As you are the one responsible."

He also stood up. His eyes widened and he now looked troubled. I acted as if I was not in the least bit concerned.

"Now if you will excuse us Mister __________ we've had a VERY long night and we need to be going for our rest at the hotel Radisson Blu. It was very nice meeting you." I turned to leave, but he had found his voice at last.

"WAIT Captain James! I think you will be able to take the instruments with you! We just need to have a list of what is being brought into the country and then you can take them, PLEASE SIR...?"

A mere five minutes later the whole crew were pushing all of our trolleys including the instruments, out the front of the terminal and across the hot carpark to the waiting bus. We were all smiling then. I looked up at the sky, *what a lovely day it was in Rajasthan today.*

Self & Cpt Doug Knights – Namaste!
In the lobby of the Radisson Blu, Jaipur (slightly delirious)

GLOSSARY OF TERMS AND ABBREVIATIONS.

AAIB – Air Accident Investigation Branch (of Dept for Transport UK)

ABP – Able Bodied Passenger

ACMI – Aircraft, Crew, Maintenance and Insurance – also known as "Wet Lease" (i.e. With Crew)

ADD – Acceptable Deferred Defect (also called "Hold Item")

ADF – Auto Direction Finder unit – often associated with NDB see below

AEW – Airborne Early Warning

AFDS – Automatic Flight Director System

AGL – Above Ground Level

AI – Attitude Indicator or Artificial Intelligence

ANO – Air Navigation Order

APU – Auxiliary Power Unit – provides electrics and hydraulics for the aircraft

AOC – Air Operators Certificate

ASR – Air Safety Report

ASI – Air Speed Indicator

ATA – Actual Time of Arrival

ATC – Air Traffic Control

ATC Slot – See Slot

ATD – Actual Time of Departure

ATIS – Automatic Terminal Information System gives aerodrome current weather – I was in love with the Geneva Volmet girl (126.80 MHz) until I realised her voice was a computer... :-(

ATL – Aircraft Technical Log or "the Techlog" (RAF Form 700)

ATO – Approved Training Organisation

Autobrake – as it says, an automatic braking system for landing

A/C – shorthand for Aircraft, can also be AC or a/c

BAe – British Aerospace

Baro – Barometric Altimeter

BBJ – Boeing Business Jet – Biz Jets based on Boeing airliners, they come in several different sizes and configurations, none of them cheap

Blocks – as in "Off Blocks" – literally means the chocks holding the wheels – same as "Off Chocks"

BER – Beyond Economic Repair

BSI – Borescope Inspection of engine

CAA – Civil Aviation Authority

CAPT – Captain or Aircraft Commander

CAS – Calibrated Air Speed = IAS corrected for instrument and position error

CAVOK – Cloud And Visibility OK – very nice weather

CB – Cumulonimbus convective build up (storm) clouds

CB – Circuit Breaker

CBT – Computer Based Training

CDL – Configuration Deviation List

CDU – Computer Display Unit

CFP – Computer Flight Plan

CFIT – Controlled Flight Into Terrain

CGI – Chief Ground Instructor

Chocks – large blocks which are stuck next to the wheels on the ramp – stops the aircraft moving if the parking brake fails

CIP – Commercially Important Passenger – same as VIP

CLB – Climb abbreviation for FMC

Coffin Corner – as the name implies, a rather undesirable place – the small corner of the flight envelope between slow speed stalling speed and high-speed buffet speed

CONFIG – Configuration of the aircraft, flaps, gear, speedbrakes etc.

Configuration Warning – warning that CONFIG incorrect for manoeuvre

CPDLC – Controller-Pilot Data Link Communications

CPT – Captain - alternative abbreviation

CRM – Crew Resource Management

CRMI – Crew Resource Management Instructor (a facilitator really)

CRS – Certificate of Release to Service (maintenance term)

CRZ – Cruise abbreviation for FMC

CRZ ALT – Cruise Altitude abbreviation for FMC

CSI – Combat Survival Instructor

CSR – Cabin Safety Report

CTBL – Contactable – referring to a crew member being available to call by crewing

CTC – Chief Training Captain (usually Postholder for Training on behalf of CAA)

CTOT – Calculated TakeOff Time – used on Flightplans

CVR – Cockpit Voice Recorder

Cyan – Colour of blue specific to Boeing EFIS displays

Damp Lease – an ACMI agreement with cockpit crew only

DDG – Dispatch Deviation Guide

DEC – Direct Entry Captain

DFDR – Digital Flight Data Recorder

DFO – Director Flight Operations – usually the Postholder for Flt Ops

DOB – Death on Board

DODAR – Diagnose, Options, Decision, Action, Review – a Decision Making tool used by some airlines

DV Window – Direct Vision window in flightdeck which slides open on ground

EAT – Estimated Approach Time (normally for holding aircraft in the stack)

EADI – Electronic Attitude Director Unit (part of EFIS instrumentation in Boeings)

EET – Estimated Elapsed Time (enroute on a Flight Plan)

EFIS – Electronic Flight Instrumentation System

EFATO – Engine Failure After Takeoff

EFOTO – Engine Failure On Takeoff

EHSI – Electronic Horizontal Situation Indicator

EMA – East Midlands Airport (IATA Code)

EOBT – Estimated Off Blocks Time

EPR – Engine Pressure Ratio – an expression of how much thrust is being produced

ETA - Estimated Time of Arrival

ETOPS – Extended-range Twin-engine Operational Performance Standards – often over water (e.g. Atlantic)

FCOM – Flight Crew Operations Manual

FCTM – Flight Crew Training Manual

FCU - Fuel Control Unit

FDAU – Flight Data Acquisition Unit – feeds the QAR

FDM – Flight Data Monitoring

FDR – Flight Data Recorder

FFS – Full Flight Simulator

FGS – Flyglobespan – an airline with HQ in Scotland 2003-2009

Final Line Check – After Line Training is complete to release crew member to Line

Fire Handle – shuts off fuel, hydraulics and when rotated activates Fire Extinguisher

Flightdeck – the little room at the front where the pilots sit

Flight Level – or FL – an altimeter indication in 000's of feet referenced to 1013mb

FMC – Flight Management Computer

FO – First Officer also known as Co-Pilot

FOD – another version of Flight Operations Director

FOD – Foreign Object Debris

FOM – Flight Operations Manager – usually Postholder reporting to the CAA

FOM – Flight Operations Manual

FORDEC – Facts, Options, Risks, Decide, Execute, Check – another Decision making acronym used by some airlines

FORM 700 – The Techlog for an RAF aircraft

FPL – Flightplan

FSTD – Flight Simulator Training Device

FSO – Flight Safety Officer

FTL – Flight Time Limitations

FUBAR – F***** Up Beyond All Recognition (MIL)

GAFA – slang for the Great Australian F*** All – central Australia

G/A – Go-around, aborted landing. (Used to be known as Overshoot)

GENDEC – General Declaration Form for Customs with all the crew names

Glass Cockpit – refers to the introduction of computer screens in flightdecks EFIS

GPU – Ground Power Unit – provides electrics to run the aircraft without APU

GPWS – Ground Proximity Warning System (does what it says on the tin)

GRADE – Gather (info), Review, Analyse, Decide (& Do), Evaluate

Ground Lock Pins – heavy duty bolts with flags to lock landing gear down

Ground Service Bus – allows the aircraft mains electrics to be powered for ground servicing operations like airline cleaners with vacuum cleaners

HEA – High Energy Approach (to land)

HF – High Frequency – Military Radio Network for Mil A/C

HSI – Horizontal Situation Indicator (Orville and Wilbur called it a compass)

Hold Item – like ADD. Holding over until repair is possible

HOTAC – Airline speak for crew "Hotel Accommodation"

IAS – Indicated Air Speed

IFALPA = International Federation of Air Line Pilots' Associations.

IFR – Instrument Flight Rules

ILS – Instrument Landing System – for precision approaches to runway in IMC

IMC – Instrument Meteorological Conditions – cloudy outside

IR – Instrument Rating – a licence to fly aircraft on instruments alone

IRS – Inertial Reference System

IRU – Inertial Reference Unit

ITCZ – Inter Tropical Convergence Zone

KORA – Key Operational Risk Area – Ryanair description

LCA – Low Cost Airline

LCC – Low Cost Company

LCZR – Localizer (element of the ILS)

Line Check – annual event for all crew

Line Training – after basic/initial training in Simulator, training on the line

LLZ – Localizer (azimuth) element of the ILS

LNAV – Lateral Navigation system – part of the AFDS

LOC – Another abbreviation of Localizer

LOCI – Loss Of Control Inflight

LOFT – Line Oriented Flying Training

LPC – Licence Proficiency Check (for pilots)

LST – Licence Skills Test (for the issue of a Type Rating)

LTC – Line Training Captain

LTCC – Line Training and Checking Captain

LTN – Luton Airport (3 letter IATA code)

LVPs – Low Visibility Procedures

LVOs – Low Visibility Operations

Mach No – associated with cruise speed of airliner as percentage of Mach 1. For example 'Mach 0.8' = 80% of Mach 1.

Magenta – purple type colour specific to Boeing EFIS system

MAN – Manchester Airport (3 letter IATA code)

MEL – Minimum Equipment List

METAR – Meteorological Actual Report – "the latest weather"

MHz – Megahertz – the VHF frequency we use for ATC

MPA – Maximum Power Assurance engine runs – Engineering Technical

MPL – Multi Pilot Licence

MOR – Mandatory Occurrence Report – to CAA (or regulating authority)

MSA – Minimum Safe Altitude

MSD – Minimum Separation Distance (used by the Military to describe closest distance to ground or other surface obstacles)

N1 – engine gauge showing speed of front fan of large bypass engines

N2 – second stage fan

N3 – third stage fan

NDB – Non-Directional Beacon – used with aircraft auto direction finder

NG – Next Generation, a Boeing term for advanced EFIS display aircraft

NMC – Network Management Centre (Ops Room)

NPA – Non-Precision Approach on instruments to an airfield/aerodrome

OAT – Outside Air Temperature – also known as Ambient

OOH – Out Of Hours for the next required Flight Duty Period

OPC – Operator Proficiency Check for crew

P1 – Senior Pilot onboard – another name for Captain – refers to logging the flight-time

P1/S – First Pilot under Supervision/Training

P2 – Co-pilot

PA – Public Address system – aircraft loudspeaker system to cabin

PAPIs – Precision Approach Path Indicators – lights at the runway threshold showing visually if you are high or low on the 3-degree glideslope

PF – Pilot Flying (also known as Handling Pilot – with hands on controls)

PIC – Pilot In Command

PAP – Passenger x 1

PAX – Passengers Plural

PIOSEE – Problem, Information, Options, Select, Execute, Evaluate – a decision making acronym favoured by Ryanair

PIREP – Pilot Report – usually of meteorological phenomena

PLI – Pitch Limit Indicators

PLOC – Prolonged Loss Of Communication (on the radio) – due to losing contact with ATC

PRM – Person of Reduced Mobility

Pusser – Generic Term for Royal Naval Supply Branch

QAR – Quick Action Recorder, receives >2000 parameters from the FDAU

QFI – Qualified Flying Instructor

QRH – Quick Reference Handbook – with emergency & non-normal checklists

Quarter-Mil – a topographical Map with scale 1:250,000

QWI – Qualified Weapons Instructor

RA – Radio Altimeter

RADALT – Radio Altimeter

RHIP – Rank Has Its privileges – a saying originally from the Army

Rotate – Call by Monitoring Pilot to indicate to PF that it is speed to get airborne

RPM – Revolutions Per Minute – with reference to engine speed usually

R/T – Radio Telephony – often abbreviated – VHF in Civil World

RTO – Rejected Takeoff

SAS – Special Air Services

SBY – Standby – referring to crew usually associated with rostering/crewing dept.

SCCM – Senior Cabin Crew Member

SID – Standard Instrument Departure – defined for all flights heading that direction

Slot – An ATC timed departure restriction – usually with an allowance of -5 and +10 minutes

SMS – Safety Management System

SNAFU – Situation Normal, All F***** Up (MIL)

SOP – Standard Operating Procedure

Soup Dragon – Cabin crew in forward galley serving flightdeck

SP – Safety Pilot – qualified observer on jumpseat in flightdeck for Line Training

Speedbird – British Airways' R/T callsign

SSA – Sector Safe Altitude – usually within 25 nm

STA – Scheduled Time of Arrival

STAR – Standard Arrival Routing – defined for all flights from that direction

STD – Scheduled Time of Departure

'Stovie' – RN Fleet Air Arm slang for 'Stovepipe Driver' – fixed wing jet pilot

TAF – Terminal Aerodrome Forecast – weather forecast at an airfield

TAI – Thermal Anti-Ice for the engines on a jet airliner

TAS – True Air Speed = speed of a/c relative to the airmass

TCAS – Traffic Collision Avoidance System

Techlog – aircraft Technical Log; a legal document – see also ATL

TEM – Threat and Error Management

Thrust Reverser – on each engine deflects exhaust forward to reduce speed on landing

Topple-Free – Gyro Attitude Indicator immune to toppling

TRE – Type Rating Examiner, check airman, usually a Training Captain

TRI – Type Rating Instructor, usually a Training Captain

TRTO – Type Rating Training Organisation

TS – Thunderstorm abbreviation in weather forecast

TWI – Tactical Weapons Instructor

TWU – Tactical Weapons Unit

UNMIN – Unaccompanied Minor – an escorted child passenger with a large tag on a lanyard round his/her neck

UTC – Universal Time Coordinated (used to be known correctly as GMT)

V1 – Decision Speed on takeoff roll beyond which we must fly

VFR – Visual Flight Rules

VHF – Very High Frequency radio – civil airliners use it

VIP – Very Important Person – many Bizjet pax are VIP (see CIP)

VMC – Visual Meteorological Conditions

VNAV – Vertical Navigation System of FMC for automatic flight

VNE – Velocity Never Exceed

VOLMET – VHF Met reports automatically broadcast – Originally from the French VOL = Flight & MET = Meteo

VOR – VHF Omnidirectional Radio Beacon

VSI – Vertical Speed Indicator – in feet per minute for climb or descent

VVIP – Usually applies to Head of State, Royalty or Head of Government (Prime Minister for example)

WAI – Wing Anti-Ice system for a jet airliner

Wet Lease – an aircraft leased to another operator WITH crew

WTL – Worn To Limits – engineering technical

WX - Weather

ABOUT THE AUTHOR

James McBride - born in Chester, UK, the eldest of four brothers. He always wanted to fly aeroplanes for a living since he was a little boy. On leaving school he worked as a nurse, and then had a career change in his twenties.

McBride joined the Royal Navy to fly jets and flew the Harrier and helicopters before going civil in the late 1980's.

He has been an airline Training Captain since 1998 and in 2014 wrote a book on aviation called *The Flightdeck Survival Manual* – 'How to survive a career flying aeroplanes for a living'. Then in 2016 he published a sequel called, *From the Flightdeck* – 'More tales from the sharp end', also available on Amazon.

As the owner of a ski chalet (Chalet Louis) in the French Alps for 10 years, from 1990 he ran skiing holidays for family and friends. In 2015 he wrote a book called *Cooking On The Piste* – 'A Ski Chalet Cookbook' with his friend Mark Chetham.

In 2018, he co-authored another aviation book called *AirCrew Stories Volume 1* – 'Fascinating real life stories from the exhilarating world of flying', with his friend Kais Kassim, available on Amazon.

He has written a regular monthly column for UK publication Flight Training News since 2006. His writing has focused on flight safety and training. He was short-listed for an Aerospace Journalist of the Year Award (AJOYA) in Paris in 2007.

Married to Rosemary, they spend their time between the USA and Europe. Hobbies include Sailing, Diving, Motorcycling, Cooking, Skiing, Watersports and Travel.

Made in the USA
Monee, IL
04 November 2026

67386433R00134

Acid and Sour Gas Treating Processes

Gulf Publishing Company
Book Division
Houston, London, Paris, Tokyo